Adsorption and Ion Exchange—'83

Y. H. Ma, D. O. Cooney and A. L. Hines, editors

...aham
...own
...ritten
...Clemens
...Cooney

...A. Hasanain
...ayhurst
...Hines
...ang

S. M. Klein
Jeng Cheng Lee
Alexander P. Mathews
Alan L. Myers
Robert P. O'Brien
John D. Y. Ou
Wayne G. Schuliger
Bryan J. Travis
Nien-Hwa Linda Wang
T. Y. Yan

AIChE Symposium Series

Number 230　　　1983　　　Volume 79

Published by
American Institute of Chemical Engineers

345 East 47 Street　　　New York, New York 10017

Copyright 1983

American Institute of Chemical Engineers
345 East 47 Street, New York, N.Y. 10017

AIChE shall not be responsible for statements or opinions advanced in papers or printed in its publications.

Library of Congress Cataloging in Publication Data

Main entry under title:
  Adsorption and iron exchange '83.

  (AIChE symposium series; no. 230)
  "Papers from sessions organized by the Adsorption and Ion Exchange Committee of Group 2 (Diffusional Operations and Processes) and held at the 1982 Annual Meeting (November, 1982) in Los Angeles, 1983 Spring Meeting (March, 1983) in Houston, and 1983 Summer Meeting (August, 1983) in Denver"—Foreword.
  1. Adsorption—Congresses. 2. Ion exchange—Congresses. I. Ma, Y. H., 1936-   . II. American Institute of Chemical Engineers. Adsorption and Ion Exchange Committee. III. Series.

QD547.A364 1984   541.3′453   83-25814
ISBN 0-8169-0267-4

FOREWORD

This volume contains papers from sessions organized by the Adsorption and Ion Exchange Committee of Group 2 (Diffusional Operations and Processes) and held at the 1982 Annual Meeting (November, 1982) in Los Angeles, 1983 Spring Meeting (March, 1983) in Houston and 1983 Summer Meeting (August, 1983) in Denver.

The papers in this volume represent both recent theoretical and experimental developments in adsorption and ion exchange. It is the intent of the Committee to publish symposium series volumes regularly to report recent research in this important area.

I would like to express my appreciation to D. O. Cooney for his assistance in selecting the fine papers for presentation in the sessions and inclusion in this volume. I also would like to thank C. Chi, A. L. Hines and the late Theodore Vermeulen, co-chairmen of the sessions, for their assistance in organizing same. In addition, I would like to thank the authors and the reviewers who enabled this volume to be assembled and to the national office of the AIChE for its very efficient handling of the publication process.

Yi Hua Ma
Worcester Polytechnic Institute
Worcester, Massachusetts 01609

CONTENTS

FOREWORD iii

ADSORPTION FROM GAS MIXTURES ON HETEROGENEOUS SURFACES .... Alan L. Myers and John D. Y. Ou ... 1

CALCULATION OF SURFACE SITE-ENERGY DISTRIBUTIONS FROM ADSORPTION ISOTHERMS AND TEMPERATURE-PROGRAMMED DESORPTION DATA .... Jerald A. Britten, Bryan J. Travis and Lee F. Brown ... 7

ADSORPTION IN AN AGITATED SLURRY OF POLYDISPERSE PARTICLES .... Alexander P. Mathews ... 18

MULTICOMPONENT ION EXCHANGE IN FIXED BEDS FOR SELECTIVE REMOVAL OF AMMONIUM CARBONATE .... Nien-Hwa Linda Wang and Stanley Huang ... 26

RECOVERY OF URANIUM FROM DILUTE SOLUTION: A NEW APPROACH .... T. Y. Yan ... 36

TREATMENT OF CONTAMINATED GROUNDWATER WITH GRANULAR ACTIVATED CARBON .... Robert P. O'Brien, Malcolm M. Clemens and Wayne G. Schuliger ... 44

ADSORPTION OF ETHANOL AND WATER VAPORS BY SILICALITE .... S. M. Klein and W. H. Abraham ... 53

ADSORPTION KINETICS FOR SYSTEMS THAT EXHIBIT NONLINEAR EQUILIBRIUM ISOTHERMS .... Mohammed A. Hasanain, Anthony L. Hines and David O. Cooney ... 60

HIGH PRESSURE OXYGEN, NITROGEN AND ARGON ADSORPTION IN MORDENITES .... David T. Hayhurst and Jeng Cheng Lee ... 67

COMPARISON OF PERFORMANCE OF PACKED AND SEMIFLUIDIZED BEDS FOR ADSORPTION OF TRACE ORGANICS .... Alexander P. Mathews and L. T. Fan ... 79

ADSORPTION FROM GAS MIXTURES ON HETEROGENEOUS SURFACES

ALAN L. MYERS
and
JOHN D. Y. OU

Department of Chemical Engineering
University of Pennsylvania
Philadelphia, Pennsylvania

This work was stimulated by recent observations of negative deviations from Raoult's law in adsorbed mixtures, especially at high coverage approaching saturation. These activity coefficients are calculated for the <u>average</u> composition of the adsorbed solution, a source of error because the actual composition varies on a heterogeneous surface. The objective is to develop a heterogeneous extension of the homogeneous theory of ideal adsorbed solutions (IAS)

INTRODUCTION

Recently, Reich, Ziegler and Rogers [3] reported data on adsorption of methane, ethane and ethylene gases, as well as their binary and ternary mixtures, on activated carbon. Three isotherms at 212.7, 260.2 and 301.4 K were measured. They found that the selectivity ($s = x_1 y_2 / x_2 y_1$) decreased rapidly, almost exponentially, with increasing pressure. For example, Figure 1 shows the selectivity of ethane(1) relative to methane(2) at 301.4 K. In one decade of pressure change (100 to 1000 kPa) the selectivity falls from about 15 to 5. At pressures below 100 kPa, the selectivity increases so rapidly that the limit at zero pressure cannot be determined by extrapolation. Figure 2 shows the isobaric variation of selectivity with composition. The typical maximum in selectivity for mixtures dilute in the heavy component (ethane) is observed.

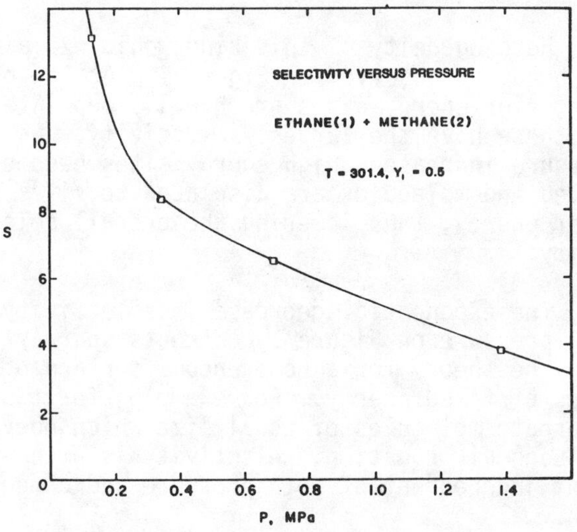

These results can be explained in terms of the heterogeneity of the activated carbon. Consider, for example, a surface composed of two types of sites: "A" sites of high energy and "B" sites of low energy. Suppose that adsorbate #1 has twice the energy of adsorption of adsorbate #2, on both sites. On each site the selectivity is equal to the ratio of

---

John D.Y. Ou is now with UOP Process Division, Des Plaines, Illinois, 60016.

Figure 1. Experimental data [3] for adsorption of binary mixtures of ethane + methane on BPL activated carbon at 301.4 K, equimolar vapor.

Henry's constants. But Henry's constants, or adsorption second virial coefficients, are exponential functions of the energy of adsorption. Therefore, because the ratio of energies of adsorption is identical on both sites, the ratio of Henry's constants (i.e. selectivity) increases exponentially with the energy of the site.

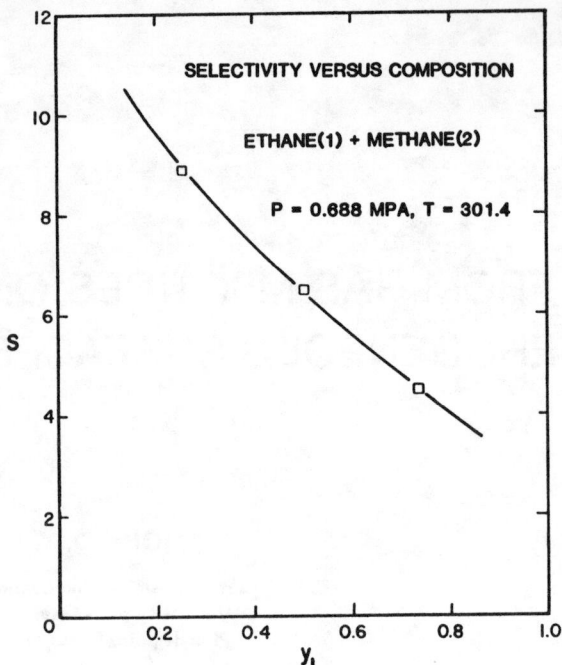

Figure 2. Experimental data [3] for adsorption of ethane + methane on BPL activated carbon at 301.4 K, 0.688 MPa.

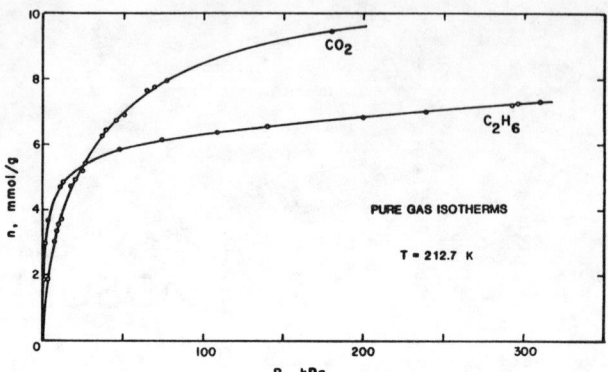

Figure 3. Pure-vapor isotherms of ethane and carbon dioxide on BPL activated carbon at 212.7 K [3].

Heterogeneity of this kind would explain the behavior observed on Figure 1. At low pressure, high energy sites are heavily populated and these have the largest selectivity. As pressure increases, high energy sites become filled and molecules are displaced to sites of lower energy, thus lowering the overall selectivity.

The exponential decrease in selectivity with pressure on Figure 1 contrasts sharply with the theory for a homogeneous surface of sites of fixed energy. For example, for two adsorbate molecules of equal size which obey the Langmuir equation, selectivity is a constant independent of both pressure and composition.

In addition to heterogeneity, there is a secondary but important size effect apparent on Figure 3. The pure-vapor adsorption isotherms cross one another. Surprisingly, this behavior is typical. The reason is that the larger molecule (here, ethane) has a smaller saturation capacity m:

$$m = V/v^{sat} \qquad (1)$$

where V is the volume of the pores per unit mass of adsorbent, and v^{sat} is the molar volume of saturated liquid. This approximation is called Gurvitch's rule. The larger molecule is usually preferentially adsorbed at low coverage but the smaller molecule has a higher saturation capacity. This crossover effect always causes the selectivity to decrease with increasing pressure. If the two adsorbates have similar values of Henry's constants (this is not the case in Figure 3) then the decrease of selectivity with pressure can lead to an adsorption azeotrope at pressures greater than the crossover point.

In summary, both effects, heterogeneity and size, cause the selectivity to decrease with increasing pressure. The heterogeneity is most important at low coverage, but at high coverage the size effect can dominate. The objective is this work is to derive equations for adsorption on a heterogeneous surface for the usual case of molecules of different size.

BETA DISTRIBUTION OF ENERGIES

The energy distribution for a heterogeneous adsorbent may be derived from the pure-vapor adsorption isotherm by applying the Beta-function equation [2]:

$$\theta = \left[1 + \frac{1}{(a-1)^{a-1}}\left[\frac{-\ln x}{b}\right]^a\right]^{-b} \qquad (2)$$

where $\theta = n/m$ is the fractional filling of micropores, $x = P/P^s$ is the dimensionless pressure, n is moles of gas adsorbed per unit mass of adsorbent, and m is the value of n at saturation ($P = P^s$ or $x = 1$).

Values of the constants a and b are constrained by the equations:

$$\frac{B_{1s} P^s}{mRT} = \left(\frac{e^{a-1}}{a}\right)^b \qquad (3)$$

$$\frac{-\Delta G}{mRT} = \frac{b}{a}(a-1)^{1-\frac{1}{a}} \left[\frac{\Gamma(b - \frac{1}{a})\Gamma(\frac{1}{a})}{\Gamma(b)}\right] \qquad (4)$$

The constants a and b, as well as the groups on the left-hand-side of Equations (3) and (4), are dimensionless. Γ is the gamma function, B_{1s} is the adsorption second virial coefficient, and ΔG is the free energy of immersion of the adsorbent in bulk liquid adsorbate.

There are four constants in Equation (2): a, b, m and P^s. The vapor pressure P^s is of course determined by the properties of the bulk liquid. The constant a is fixed for a particular adsorbent, and is equal to 2 for most commercial brands of activated carbon. The saturation capacity is found experimentally, or it may be estimated from the pore volume of the adsorbent by Equation (1). This leaves one constant, b, which is determined experimentally.

Integration of Equation (2) for the spreading pressure leads to an incomplete Beta function, which because of its relation to the Hypergeometric function may be expressed as an infinite series:

$$\left[\frac{\Pi A}{-\Delta G}\right] = 1 - \frac{t^{\frac{1}{a}}}{B\left(1+\frac{1}{a}, b-\frac{1}{a}\right)} \left\{\frac{1}{b}(1-t)^{b-\frac{1}{a}} \right.$$
$$\left. + \frac{t}{(1+\frac{1}{a})} \sum_{n=0}^{\infty} \frac{(1+\frac{1}{a})_n (1+\frac{1}{a}-b)_n}{(2+\frac{1}{a})_n} \frac{t^n}{n!}\right\} \qquad (5)$$

where

$$t = 1 - \theta^{\frac{1}{b}} \qquad (6)$$

$$(m)_n = \frac{\Gamma(m+n)}{\Gamma(m)} \qquad (7)$$

$$B(m,n) = \frac{\Gamma(m)\Gamma(n)}{\Gamma(m+n)} \qquad (8)$$

B is the Beta function. Equation (5) converges rapidly except at very small values of θ.

The local step isotherm approximation upon which Equation (2) is based relates the adsorption isotherm to the energy distribution by the equation:

$$z = -RT \ln x = -RT \ln(P^*/P^s) \qquad (9)$$

P^* is the pressure at which condensation occurs on a site with energy of adsorption z relative to saturated liquid adsorbate. The step adsorption isotherm and its thermodynamic properties are summarized in Table 1.

Table 1. Thermodynamic Properties of Step Adsorption Isotherm.

Pressure Range	Amount Adsorbed	Spreading Pressure, $\Pi A/m$
$P < P^*$	0	0
$P = P^*$	m	0
$P^* < P < P^s$	m	$RT \ln(P/P^*)$
$P = P^s$	m	$RT \ln(P^s/P^*) = z$

Combination of Equations (2) and (9) gives the energy of each site corresponding to fractional pore filling:

$$\frac{z}{RT} = b(a-1)^{\frac{a-1}{a}} \left(\theta^{-\frac{1}{b}} - 1\right)^{\frac{1}{a}} \qquad (10)$$

In Table 2 are given the constants of the Beta function equation for adsorption of ethane and carbon dioxide at 212.7 K, corresponding to the experimental data on Figure 3.

MIXED-GAS ADSORPTION, HOMOGENEOUS SURFACE

Before consideration of a heterogeneous surface, the equations of equilibrium for a homogeneous surface are summarized. If the gas phase is perfect and the adsorbed solution obeys Raoult's law:

$$Py_1 = P_1^o x_1 \qquad (11)$$

$$Py_2 = P_2^o x_2 \qquad (12)$$

$$\frac{1}{n} = \frac{x_1}{n_1^\circ} + \frac{x_2}{n_2^\circ} \qquad (13)$$

There are three degrees of freedom for binary adsorption (T,P,y_1). The quantities n_1° and P_1° refer to moles adsorbed and pressure for the pure gases adsorbed at spreading pressure Π and temperature T. Since $x_2 = (1 - x_1)$, the above set of three equations, which is called ideal adsorbed solution theory (IAST), can be solved for the three unknowns (Π, n, x_1). For the Beta function isotherm, Π is given by Equation (5).

MIXED-GAS ADSORPTION, HETEROGENEOUS SURFACE

In order to extend Equations (11)-(13) to the case of a heterogeneous surface, some assumptions are introduced. First, the concept of an ideal adsorbed solution is retained but it is applied to individual sites instead of the entire surface. Second, it is assumed that the local adsorption isotherm for a particular energy may be approximated by a step function. This is the same assumption used to derive Equation (2).

It is necessary to match sites of different vapors. In the case of carbon, the success of Equation (1) implies that corresponding "sites" are actually micropores of a particular size in which the energy of adsorption is relatively constant. Thus sites should be matched at equal values of θ:

$$\theta_1 = \theta_2 \qquad (14)$$

from which values of z and P^* may be calculated for both vapors on each site. The equations of equilibrium for a particular site are:

$$Py_1 = P_1^* \exp(\Pi A/m_1 RT) x_1 \qquad (15)$$

$$Py_2 = P_2^* \exp(\Pi A/m_2 RT) x_2 \qquad (16)$$

The exponential term in these equations is a kind of Poynting correction to ensure thermodynamic consistency. This factor is not in the equations for a homogeneous surface because the standard states are at the same spreading pressure as that of the mixture. For vapor-liquid equilibria, the Poynting correction can usually be ignored but here it is important. At the standard states P_1^* and P_2^* the spreading pressure is zero (see Table 1). Using Equation (9) the above equations may be written:

$$Py_1 = P_1^s \exp(\Pi A/m_1 RT) \exp(-z_1/RT) x_1 \qquad (17)$$

$$Py_2 = P_2^s \exp(\Pi A/m_2 RT) \exp(-z_2/RT) x_2 \qquad (18)$$

Thus the size effect mentioned previously enters into the equations of equilibrium by means of the Poynting correction. If the molecules are of equal size $(m_1 = m_2)$, it follows that the selectivity of a site is given by:

$$s = (P_2^s/P_1^s) \exp\left\{\frac{z_1 - z_2}{RT}\right\} \qquad (19)$$

Thus the source of the exponential variation of selectivity with coverage is apparent.

Having fixed the independent variables (T, P, y_1) and using matched values of z_1 and z_2 from Equations (10) and (14), the equilibrium Equations (17) and (18) can be solved for the unknowns Π and x_1 for any site. The local differential amount adsorbed is then:

$$n\,d\theta = \frac{1}{\frac{x_1}{m_1} + \frac{x_2}{m_2}} \qquad (20)$$

Then the overall or global adsorption isotherm is calculating by integrating with respect to θ starting with sites of highest energy and continuing to the lowest-energy sites occupied (θ°):

$$n_t = \int_0^{\theta^\circ} n\,d\theta \qquad (21)$$

$$n_{1t} = \int_0^{\theta^\circ} x_1 n\,d\theta \qquad (22)$$

$$x_{1t} = n_{1t}/n_t \qquad (23)$$

Only the global quantities, from which the overall selectivity may be calculated, can be compared with experiment. The limit of integration θ° is set by the requirement that $\Pi = 0$ on the site of lowest energy occupied (see Table 1). Since the equilibria on local sites are thermodynamically consistent, it follows that predictions using Equations (21)-(23) are also consistent.

SAMPLE CALCULATIONS

The selectivity $(x_{1t}y_2/x_{2t}y_1)$ calculated by these equations for the mixture ethane and carbon dioxide at 212.7 K is shown on Figures 4 and 5. The pure-vapor isotherms at the same temperature are shown on Figure 3, and the constants derived for the Beta function isotherm are given in Table 2. Experimental data are not available for this system, but Figure 4 predicts an exponential drop in selectivity with pressure. The limiting value of selectivity at zero pressure (570) is given by the ratio of adsorption second virial coefficients.

Table 2. Constants of Beta Function Isotherm for Adsorption on BPL Activated Carbon at 212.7 K.

Vapor	a	b	m mmol/g	P^s kPa	$\frac{-\Delta G}{mRT}$	B_{1s} cm^3/g
C_2H_6	2	34.50	6.61	373	5.263	1.18×10^6
CO_2	2	12.93	10.03	437	3.283	2.14×10^3

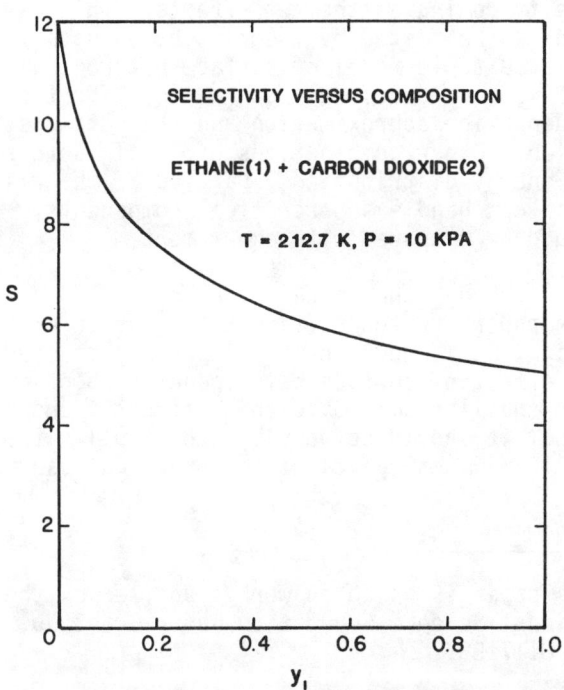

Figure 5. Prediction of selectivity of adsorbed mixtures of ethane and carbon dioxide on BPL activated carbon at 212.7 K, 10 kPa.

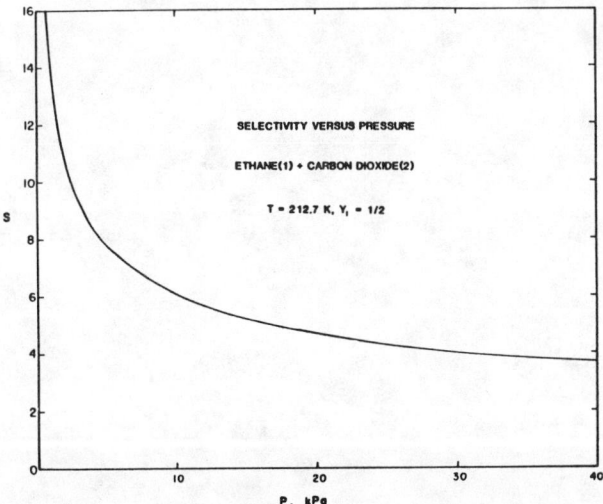

Figure 4. Prediction of selectivity of adsorbed mixtures of ethane and carbon dioxide on BPL activated carbon at 212.7 K, equimolar vapor.

DISCUSSION

Mixed-gas adsorbate-vapor equilibrium calculations for a heterogeneous surface are straightforward but time-consuming because an integral must be evaluated for each equilibrium point. The theory described here is based upon the condensation approximation, which does not apply if the temperature is above the critical temperature of either component. The results of Reich, Ziegler and Rogers [3] are for the systems methane-ethane, methane-ethylene, and ethane-ethylene. For the first two mixtures, one component is supercritical and for the last mixture, the components are too nearly alike. Unfortunately, no data were taken on the carbon dioxide-ethane mixture shown on Figures 4 and 5. However, the strong variation of selectivity with both pressure and composition is the same as found experimentally for the methane-ethane mixture shown on Figures 1 and 2.

CONCLUSIONS

Although both adsorbate-adsorbate interactions and surface heterogeneity contribute to mixture behavior, the evidence is that heterogeneity dominates the two effects. This assertion can be tested by theories based upon IAS and a suitable model of surface heterogeneity. Here we have developed such a model for the condensation approximation and the Beta distribution. Other combinations of local isotherms and energy distributions, particular theories which can handle supercritical components, should be compared with experiment.

Although the calculations described in this paper are too difficult for use in process design, it is nevertheless important to study the effect of surface heterogeneity upon adsorption equilibrium. Once this effect is understood, it should be possible to simplify the equilibrium calculations for practical use.

ACKNOWLEDGMENT

This paper is based on work supported by the National Science Foundation under Grant No. CPE-8117188.

LITERATURE CITED

1. Myers, A.L. and D.Y. Ou, "The Beta Function Adsorption Isotherm for Adsorption of Vapors on a Heterogeneous Surface," presented November 11, 1981, AIChE Annual Meeting, New Orleans, Louisiana.

2. Myers, A.L. and S. Sircar, "Principle of Correspondence for Adsorption of Vapors on Heterogeneous Adsorbents", AIChE Journal, in press.

3. Reich, R., W.T. Ziegler and K.A. Rogers, "Adsorption of Methane, Ethane, and Ethylene Gases and Their Binary and Ternary Mixtures and Carbon Dioxide on Activated Carbon", Ind. Eng. Chem. Process Des. Dev. $\underline{19}$, 336 (1980).

CALCULATION OF SURFACE SITE-ENERGY DISTRIBUTIONS FROM ADSORPTION ISOTHERMS AND TEMPERATURE-PROGRAMMED DESORPTION DATA

JERALD A. BRITTEN
BRYAN J. TRAVIS
and
LEE. F. BROWN
Los Alamos National Laboratory,
Los Alamos, New Mexico

A relatively new technique—regularization and its modifications—is applied to inverting the Fredholm integral equations posed when site-energy distributions are desired from adsorption isotherms or temperature-programmed desorption (TPD) data. Other attacks on these problems frequently require the data to be smoothed, sometimes by analytic approximation. An a priori assumption concerning the form of the distribution is needed in some approaches. Regularization requires no smoothing of the data or assumptions about the form of the solution. The method is shown to work very well in extracting site-energy distributions from adsorption isotherms, whatever the local-isotherm kernel. It also works very well in obtaining site-energy distributions from TPD data.

INTRODUCTION

This paper presents some results of applying a relatively new technique of inverting Fredholm integral equations of the first kind to the calculation of surface site-energy distributions (SED's) from adsorption isotherms and temperature-programmed desorption (TPD) data. Determinations of SED's from isotherms and TPD data are two examples of the inversion problem, which arises in many physical situations. This problem requires the calculation of a distribution function of a particular variable, given a two-dimensional model equation and data as a function of another variable. The two variables, one belonging to the distribution function and the other to the data, are the two dimensions of the model equation.

Consider the calculation of a surface site-energy distribution for gas sorption on a heterogeneous surface, given adsorption isotherm data as a function of pressure and an appropriate single-energy isotherm equation. The fraction of surface covered by an adsorbate on a nonuniform surface is a function of the gas-phase pressure of the adsorbate, i.e., $\theta_{gr} = \theta_{gr}(p)$. The fractional coverage at any pressure is equal to the fraction covered of any single-energy sites integrated over the range of possible energies:

$$\theta_{gr} = \int_{q_{min}}^{q_{max}} \theta(p,q)\eta(q)dq \quad . \qquad (1)$$

In this equation, θ_{gr} is the gross fraction of surface covered by the adsorbate, $\theta(p,q)$ is the fraction of sites of energy q covered at the pressure p (the single-energy adsorption isotherm) and $\eta(q)dq$ is the fraction of sites with adsorption energies between q and q + dq. In previous works in this area, the Langmuir and Hill-de Boer isotherms have been the most popular two-dimensional model relationships used for $\theta(p,q)$.

A similar situation occurs if it is desired to obtain an SED from temperature-programmed desorption data and the appropriate single-energy equation. Here the equation is

$$r(T) = \frac{d\theta_{gr}}{dT}(T)$$

$$= -\frac{A}{\beta} \int_{E_{min}}^{E_{max}} [\theta(T,E)]^n e^{-E/RT} \eta(E)dE \quad , \quad (2)$$

Jerald A. Britten is now at the University of Colorado, Boulder, Colorado.

in which $r(T)$ is the desorption rate as a function of temperature, $\theta(T,E)$ the fraction of sites of energy E covered at the temperature T, A the pre-exponential factor and E the activation energy for desorption, n the order of the desorption, and β the heating rate. The function $\eta(E)$ is the SED for this situation. Both Eqs. (1) and (2) are Fredholm integral equations of the first kind.

The degree of surface heterogeneity can have a profound effect on the sorption behavior of a surface, and quantitative knowledge of the variation of equilibrium and activation energies of sorption is essential in order to obtain quantitative knowledge of the parameters associated with this phenomenon. For example, Hutchinson et al. (1) stated that their experimental reaction rate data for the catalytic ortho-para hydrogen shift could not be modeled unless a distribution of adsorption and desorption activation energies was assumed. Knowledge of their SED could have reduced their modeling efforts drastically and verified this conclusion. In another type of situation, correct calculation of the adsorption entropy from adsorption data can be made only if the adsorption-energy distribution for the system is known (2).

PREVIOUS WORK

Equation (1) is well known, and the different individual details of it have been objects of intensive study. Thermodynamic factors (e.g., 3), the effects of homotattic and randomly-distributed surfaces (e.g., 4), how differing local isotherms influence the resulting distribution (e.g., 5), all have received significant scrutiny.

Of all the aspects of Eq. (1), inversion of it to obtain the site-energy distribution has been the the most troublesome, and this inversion has been the subject of a very large number of papers over the past forty years. There are difficulties inherent in inverting a Fredholm equation of the first kind. The problem is "ill-posed" if the kernel $\theta(p,q)$ is smoother or more continuous in the variables than is $\theta_{gr}(p)$. Such is invariably the case in adsorption, since $\theta_{gr}(p)$ contains experimental error, while $\theta(p,q)$ is analytic in p and q. Direct inversion of Eq. (1) leads to wildly oscillating solutions which depend on the number of significant figures in the input data (6). Actually, a unique classical solution to Eq. (1) does not exist, even if errors in $\theta_{gr}(p)$ are as small as computer-generated roundoff and truncation errors.

Many investigators attacked this difficulty by assuming an analytical expression for $\theta_{gr}(p)$ such that its order of smoothness matched that of $\theta(p,q)$. Various methods were then used for the inversion. Analytical techniques were developed by Temkin and Levich (7) and Sips (8,9). These investigators showed that when the Langmuir isotherm is used for $\theta(p,q)$, appropriate substitutions can convert Eq. (1) into an equation for an integral transform. These can be inverted to give $\eta(q)$ for some specific analytic approximations of $\theta_{gr}(p)$ such as the Temkin or Freundlich isotherms. Another approach was to use the "condensation approximation," in which the kernel of Eq. (1) was approximated by a step function (10). An elaboration of this involved measuring the adsorption energy as a function of coverage at different temperatures and extrapolating $(d\theta_{gr}/dq)$ to 0 K (11). At 0 K, adsorption occurs in serial order from sites of highest to lowest energy and the condensation approximation is obeyed exactly (2). Adamson and Ling (12) used the condensation approximation to acquire a first approximation to an integral SED, then used it again in an iterative scheme to improve the calculated integral SED.

Other iterative methods have been developed. Ross and Morrison (13), building upon earlier work (3), and House and Jaycock (14) treated heterogeneous surfaces as "patches" of uniform sites, each patch with a unique and constant adsorption potential. The SED in these treatments was determined by adjusting η iteratively through changing the area of each patch until the calculated isotherm agreed with the experimental data within a predetermined error limit. Van Dongen (15) chose an exponential series with adjustable parameters to approximate η, and then used nonlinear regression to obtain parameter values which minimized the square of the difference between the RHS and LHS of Eq. (1).

Rudzinski and his coworkers have developed many of these methods further in a large number of papers. Especially they have built upon the integral transform methods (e.g., 16) and the condensation approximation (e.g., 17). References to other works of this school are in these papers and more recent reviews (e.g., 18-20).

Compared with the voluminous collection of studies which analyze adsorption data on heterogeneous surfaces, little work has been done in the area of obtaining SED's from TPD data. By performing desorption experiments at differing initial surface coverages, plots of $\ln(r)$ vs. $(1/T)$ at a particular value of θ_{gr} will give as a slope the activation energy as a function of surface coverage (21). Such plots can be constructed, for the same purpose, by performing a number of desorption experiments at widely differing heating rates with the same initial surface coverage (22). This latter work also considered the least-squares minimization of an N-th order system of equations for N constants in a series approximation to the SED.

In many of the above techniques, the data cannot be used in their raw form; some sort of smoothing must be carried out first (cf. 5,18). In others, an initial approximation of the distribution (cf. 23) or some assumption as to the mathematical form of the distribution is required (cf. 15). Ross and Morrison (13) pointed out that a given isotherm can be approximated closely by several different distribution functions, and the one converged upon by an algorithm may very well depend upon the initial approximation of the function.

Another method of solving first-kind Fredholm integral equations, called regularization, recently has found favor with many workers. This method suffers from none of the disadvantages mentioned in the previous paragraph. Regularization was applied to calculating SED's from adsorption isotherms by House (4,24). In one of these papers (24), House recalculated adsorption isotherms using SED's from both regularization and the method developed by House and Jaycock (14,23). He compared these calculated isotherms with the original experimental one, and the isotherm calculated using the SED from regularization gave the poorer agreement. Perhaps because of this, more recent works by this investigator involving SED's have not used regularization (e.g., 25,26). On the other hand, Papenhuijzen and Koopal (27) compared the method of regularization with that of Ross and Morrison, and concluded that regularization was decidedly superior. There also have been some recent developments in the method of regularization. These advances, particularly the treatment of error and the employment of a nonnegativity constraint (28), have improved this method for application to problems of the type being studied here and it appears to merit further study.

In spite of the effort spent on developing methods for inverting these equations to obtain SED's, applications are few. Some of the studies mentioned above showed how the observed SED's resulted from adsorbent composition (e.g., 3,5,15), and these provide valuable insight into how adsorbent composition can affect the SED. On the other hand, no studies of how the SED affects other chemical behavior, such as reaction rate or sorption behavior, have appeared yet. Perhaps now, with the techniques for obtaining SED's approaching maturity, application is not far away.

MATHEMATICAL APPROACH

The general form of Eqs. (1) and (2) is a Fredholm integral equation of the first kind,

$$f(x) = \int_a^b K(x,t)g(t)dt \, , \quad c \leqslant x \leqslant d \, , \quad (3)$$

in which the kernel K is representative of the experimental system and procedure. The difficulties in solving this equation for the function $g(t)$ have been mentioned earlier. In any real situation, there is not a unique distribution function $g(t)$ satisfying Eq. (3), and an assumption about the distribution must be made to obtain the "physically true" solution. An approach developed by Phillips (6), Twomey (29), and Tikhonov (30-32) consisted of "regularizing" the solution by adding a smoothing term to the normal least-squares minimizing functional. Thus an assumption is made concerning the smoothness of the solution in this procedure; the greater the magnitude of the smoothing term, the smoother the solution $g(t)$ is assumed to be. Here we follow most closely the scheme of Tikhonov.

When Eq. (3) characterizes a physical system from which experimental observations of $f(x)$ have been obtained, then a solution for $g(t)$ would minimize locally the integral

$$I = \int_c^d [f_e(x) - \int_a^b K(x,t)g(t)dt]^2 dx \, , \quad (4)$$

in which $f_e(x)$ represents the experimental observations. A smoothing term is added to the RHS of Eq. (4), and the functional to be minimized is altered so that

$$I = \int_c^d [f_e(x) - \int_a^b K(x,t)g(t)dt]^2 dx + \alpha \int_a^b [H(g)]dt , \quad (5)$$

where $H(g)$ is a nonlinear differential operator on g with nonnegative coefficients, and α is a parameter. In the simplest formulation, $H(g) = g^2$, which is what we shall use throughout this work. Using the calculus of variations to obtain the necessary condition for a minimum yields

$$\alpha g(t) + \int_a^b [\int_c^d K(x,z)K(x,t)dx]g(z)dz = \int_c^d K(x,t)f_e(x)dx . \quad (6)$$

The details of the variational operations used in obtaining Eq. (6) are presented in the Appendix.

Equation (6) can be put into finite-difference form, and this gives

$$(\underline{K}^{tr} \underline{\delta x} \, \underline{K} \, \underline{\delta t} + \alpha \underline{I})\underline{g} = \underline{K}^{tr} \underline{\delta x} \, \underline{f}_e , \quad (7)$$

in which $\underline{K}^{tr}$ is the transpose of $\underline{K}$, and $\underline{\delta x}$ and $\underline{\delta t}$ are diagonal matrices whose elements are the weighting factors for the intervals $[c,d]$ and $[a,b]$. There is a unique $g(t)$ which gives a minimum to the integral in Eq. (5). As α approaches zero, Eq. (5) approaches Eq. (4), the original problem. For some value of α, an optimal solution is found. The weakness of this approach in its original form is that α is somewhat arbitrary. Sometimes the value of α that results in the smallest value of I in Eq. (4) is chosen as the optimal value. Papenhuijzen and Koopal (27) used the α that minimized $\int g''(t)dt$, which means that the solution must possess a good degree of smoothness. If α is too small, the solution begins to oscillate, and loses its physical realism.

Solution of Eq. (5) by unmodified regularization usually results in minor oscillations in $g(t)$, giving negative values for some points within the range of t. For many applications, negative values of $g(t)$ are impossible, and the solution can be restricted to nonnegative values. A recent improvement to the regularization technique allows the inclusion of a nonnegativity constraint on the distribution. Butler et al. (28) explained how this can be done in an iterative manner. They also used $H(g) = g^2$ in their work. Equation (7), using a given α, is solved for $\underline{g}$. The points at which $\underline{g}$ is negative are recorded, and the evaluation of the quantity $(\underline{K}^{tr} \underline{\delta x} \, \underline{K} \, \underline{\delta t} + \alpha \underline{I})$ is not performed at these points. A new $\underline{g}$ is obtained, and the process is repeated. The iteration is continued until no change in $\underline{g}$ is seen. The entire process is then repeated for all subsequent values of α.

We carried out many trial calculations where we postulated SED's which were used to calculate isotherms via Eq. (1). We used these isotherms to extract SED's by regularization and compared them with the postulated ones. We found that inclusion of the nonnegativity constraint through Butler et al.'s approach markedly improved the agreement between the postulated and calculated SED's.

Another, and very important, contribution was made in the paper by Butler et al. They developed a criterion, based on the error in the data, for specifying the value of the smoothing parameter α. This is discussed below when we consider the treatment of data containing error.

Holt and Jupp (33) suggested another improvement in regularization. They used spline functions to represent the solution $g(t)$, but allowed the nodes to float. The nodes were concentrated where the gradients of the distribution were large, providing high resolution of narrow peaks and near discontinuities. We have reached the same goal by allowing a variable integration mesh in our computer code. For each value of α, the integration mesh is refined to allow finer zoning in regions where the emerging distribution has large gradients.

IDEAL SYSTEMS

To check the value of regularization for inverting adsorption isotherms and TPD data, some simple SED's were conceived and used to generate isotherms and TPD spectra by solving Eqs. (1) and (2) in the forward direction. SED's were then extracted from the calculated isotherms and spectra through regularization, and compared with the postulated SED's. The nonnegativity constraint was included in the regularization code we used.

In testing regularization with adsorption isotherms, two simple SED's were used--one a unimodal, normal distribution, and the other bimodal with both peaks normally distributed. The unimodal distribution was that observed by Ross and Olivier (3) to give good results for argon adsorbed on P33(1000) graphite at 77.5 K, and the bimodal one was a simple variation on the unimodal. The generating functions were

$$\eta(q) = (1/2544)[e^{(-4.856 \cdot 10^{-7})(q-8619)^2}] \text{ joules}^{-1} \quad (8)$$

and

$$\eta(q) = (1/588)[e^{(-4.856 \cdot 10^{-7})(q-8619)^2} + e^{(-4.856 \cdot 10^{-7})(q-17238)^2}] \text{ joules}^{-1} \quad (9)$$

The first tests used a Langmuir isotherm as a kernel in the Fredholm equation. The kernel used was

$$K(p,q) = \frac{(4.464 \cdot 10^{-6})(e^{q/RT})(p)}{1 + (4.464 \cdot 10^{-6})(e^{q/RT})(p)} \quad (10)$$

The results from these tests are presented in Figs. 1 and 2. In both cases, the agreement is excellent. Actually, even better agreement could have been achieved by a more judicious choice of the isotherm-point location. Isotherms of fifty points were used. It is believed that if the points had been more strategically located, no visual distinction whatsoever could have been made between the postulated and calculated SED's.

Tests using a Hill-de Boer isotherm as the kernel were also carried out. In this case, the elements k_{ij} of the $\underline{K}$ matrix were calculated from the implicit equation

$$p_i - \frac{1}{4.464 \times 10^{-6}} \left(e^{-q_j/RT} \right) \left(\frac{k_{ij}}{1-k_{ij}} \right) \times \exp\left[\frac{k_{ij}}{1-k_{ij}} - \frac{2a_2 k_{ij}}{b_2 k_B T} \right] = 0 . \quad (11)$$

In this equation, a_2 and b_2 are the two-dimensional van der Waals constants, tabulated by de Boer (34) and by Ross and Olivier (3). We used the values for argon. The results were very similar to those observed when using the Langmuir isotherm kernel, and are not illustrated.

Similar SED's were postulated for testing the application of regularization to TPD spectra. The mean energies were much higher, for TPD normally is used for chemisorbed materials. Only the bimodal distribution is presented here, for the results obtained when using it were similar in kind to those from the unimodal one and more visible. The generating function for the bimodal distribution was

$$\eta(E) = (9.97 \cdot 10^{-4})[e^{(-1.25 \cdot 10^{-9})(E-2.5 \cdot 10^5)^2} + e^{(-1.25 \cdot 10^{-9})(E-3.0 \cdot 10^5)^2}] \text{ joules}^{-1} . \quad (12)$$

We assumed a first-order desorption, and so n = 1. In this case, the elements k_{ij} of the $\underline{K}$ matrix were equal to

$$k_{ij} = [\theta(T_i E_j)] e^{-E_j/RT_i} . \quad (13)$$

The factor $\theta(T_i E_j)$ was obtained by solving the equation

$$\frac{d\theta}{dT}(E_j) = -\frac{A}{\beta} \theta(E_j) e^{-E_j/RT} \quad (14)$$

for $\theta(E_j)$ as a function of T, and tabulating the values for the various T_i's desired. As when studying the ideal adsorption isotherms, fifty points were used when extracting the SED from the TPD spectrum.

The result of this test is shown in Fig. 3. The agreement is very good, and only minor deviations from the postulated SED are seen in the calculated one. In the equation for the TPD spectrum, the kernel is a complicated and sensitive function of both the variables, while the kernel is much less sensitive to the experimental variable in the adsorption isotherm equation. In the TPD kernel, the temperature is in the exponent; in the isotherm kernel, the pressure is not. It is possible that more points from the TPD spectrum would give the essentially perfect agreement seen when the SED's were extracted from the isotherms. Nevertheless, the agreement is still very good, matching the slopes and discriminating between two peaks with an accuracy well within any experimental error.

TREATMENT OF ERROR

There is a different degree of smoothness in the calculated $g(t)$ for each different value of the parameter α. It was pointed out earlier that as $\alpha \to 0$, the solution $g(t)$ began to fluctuate chaotically. As $\alpha \to \infty$, the solution $g(t)$ approaches a straight line, $g(t) = 0$, for all t. At some point between these two extremes, a "best" solution exists, and a criterion for choosing this "best" solution is needed. Empirical studies show that choosing the value of α which minimizes the RHS of Eq. (4) frequently yields a seriously oscillating solution unless the data are extremely accurate—sometimes more accurate than allowed by computer roundoff.

The presence of experimental error in the data is a factor in choosing the optimal solution. If data subject to error are assumed to be error-free, then structure in $g(t)$ will arise which is not really there. Butler, Reeds, and Dawson (28) have offered a criterion for choosing the optimal value of α which takes error into account. It is based on the estimated error in the problem solution, i.e., the integral of the estimated squared error in the calculated $g(t)$ over the range of t between a and b. This in turn depends on the error in the data. Their criterion may be expressed as in the following lines.

The terms are defined:

$$\underline{\underline{M}} \equiv \underline{\underline{K}}^{tr} \underline{\underline{\delta x}} \; \underline{\underline{K}} \; \underline{\delta t} \quad (15)$$

$$\underline{\underline{T}} \equiv (\underline{\underline{M}} + \alpha \underline{\underline{I}})^{-1} \quad (16)$$

$\underline{c} \equiv$ a vector which satisfies $(\underline{\underline{M}} + \alpha \underline{\underline{I}}) \; \underline{c} = \underline{f}_e$
in which $\underline{\underline{M}}$ is evaluated only for points in $\underline{t}$ where $g(t) > 0$. (17)

A function of α is then defined:

$$D(\alpha) \equiv (\underline{f}_e^{tr} \; \underline{\underline{T}} \; \underline{\underline{M}} \; \underline{\underline{T}} \; \underline{f}_e) - (2 \; \underline{f}_e^{tr} \underline{\underline{T}} \; \underline{f}_e)$$
$$+ (2\sigma \sqrt{N \; \underline{c}^{tr} \underline{c}}) \quad (18)$$

The criterion of Butler, Reeds, and Dawson (the BRD criterion) states that the optimal value of α is attained when $D(\alpha)$ is a minimum.

In their paper, Butler et al. reported the results of several tests they ran to check their criterion. They first postulated a distribution, combined it with the Laplace kernel $K = \exp(-xt)$, and used the result to generate a series of numerical data. They then imposed error on the data, and calculated $g(t)$ through regularization with the nonnegativity constraint. The results appeared to depend on the type of function being investigated. In some, the optimal α averaged about one-tenth the value specified by the BRD criterion. In another, the optimal α averaged about the criterion value.

When the levels of error in the various data points are not equal, elements in the $\underline{\underline{K}}$ matrix and $\underline{f}_e$ vector must be weighted. Let the weights w_i^2 be inversely proportional to the variances in the data points f_i. The weights are scaled so that $\sum_{i=1}^{N} w_i^2 = N$. The elements in the $\underline{\underline{K}}$ matrix are then weighted so that $k_{ij} \to w_i k_{ij}$ and those in the $\underline{f}_e$ vector so that $f_{ei} \to w_i f_{ei}$.

To test how error in adsorption isotherm data would affect the result and how the BRD criterion should be applied, we imposed a random 1% error on the data used to obtain the bimodal distribution in Fig. 2. We found that the optimal α occurred in the vicinity of one-tenth the α specified by the BRD criterion, and the resulting distribution is presented in Fig. 4 and compared with the postulated distribution.

The calculated distribution still agrees quite well with the postulated one. The location of the peaks is identified very well, and only the heights differ somewhat.

APPLICATIONS TO REAL SYSTEMS

To see if reasonable SED's can be obtained from real systems through regularization, two sets of data were analyzed. One was the adsorption of argon on sodium bromide at 77.5 K reported by Ross and Olivier (3), and the other was the TPD of CO from the surface of an oxidized graphite carried out by one of the present authors (35).

In a study of the adsorption of argon on sodium bromide at 77.5 K, Ross and Olivier used a SED which summed two normal distributions,

$$\eta(q) =$$
$$(0.25)(1/4280)[e^{(-1.71 \cdot 10^{-7})(q-6650)^2}]$$
$$+ (0.75(1/677)[e^{(-6.86 \cdot 10^{-6})(q-6230)^2}]$$
$$\text{joules}^{-1}. \qquad (19)$$

They found that this could be combined with a Hill-de Boer local-isotherm kernel to calculate an overall isotherm which gave good agreement with that observed experimentally. We took the experimental isotherm, and using the procedures described above, calculated two SED's. One was evaluated at the optimal α specified by the BRD criterion, and the second was calculated at an α of one-tenth this value. These two values of α probably are the extremes of the range which encompasses the α which will give the most probable distribution.

The two SED's are presented in Fig. 5, along with the SED of Ross and Olivier. The SED's obtained through regularization show more structure. The two distinct peaks these SED's possess differentiate more clearly between the two types of surface on the NaBr than does the postulated SED of Ross and Olivier. Nevertheless, there is excellent agreement in major characteristics between the regularization SED's and that of Ross and Olivier. The locations of the major peaks and the spreads of the distributions all coincide very well. It may be noted that the SED of Ross and Olivier was supported by other studies they made.

The SED obtained from the TPD of CO from a surface of graphite oxidized by CO_2 is presented in Fig. 6. Here it is difficult to choose an error criterion, because the data are selected from a continuous trace. Thus errors in the data probably are not random, but there may be some bias over a range of the points. Since the TPD spectrum itself is very smooth, it is not surprising that a smooth distribution is obtained. Our studies have found, however, that reproducing the original TPD spectrum from the calculated SED is much more difficult than reproducing an experimental adsorption isotherm from the calculated SED, i.e., the TPD spectrum is much more sensitive to variations in the SED than is the adsorption isotherm. Thus, the SED obtained by regularization from a TPD spectrum is much less subject to error than is the SED obtained from an isotherm.

The TPD spectrum calculated from the SED and the first-order desorption kernel is compared with the original spectrum in Fig. 7. Good agreement is seen, supporting the realism of the SED obtained.

CONCLUSIONS

Regularization, with its recent modifications to include a nonnegativity constraint and a reasonable treatment of data containing error, appears to be an excellent approach to obtaining site-energy distributions from adsorption isotherms and TPD spectra.

NOMENCLATURE

A	Frequency factor for desorption, s^{-1}
a_2	Two-dimensional van der Waals constant, J-cm^2/mol
b_2	Two-dimensional van der Waals constant, cm^2/mol
c	Nonnegative components of g in the additive smoothing term of Eq. (5), various units
D	Function of α defined by Eq. (18)
E	Activation energy for desorption, J/mol
f	Function of experimental variable, various units
f_e	Experimental observations, various units
g	Distribution function, various units
H	Nonlinear operator or nonlinear differential operator
I	An integral, various units
K	Kernel of integral equation, various units
k_B	Boltzmann constant, J/molecule-K
k_{ij}	Element of matrix K, various units
M	Matrix defined by Eq. (15)
N	Number of experimental data points
n	Order of desorption, dimensionless
p	Pressure, Pa
q	Energy of desorption, J/mol (negative of the enthalpy change upon adsorption)
R	Gas constant, J/mol-K
r	Desorption rate, s^{-1}
T	Temperature, K
T	Matrix defined by Eq. (16)
t	Variable over which distribution function g occurs, various units

w_i Weighting factor, inversely proportional to the variance in the data taken at point i

x Experimental variable, various units

z Dummy integration variable, various units

α Adjustable smoothing parameter, dimensionless

β Heating rate in temperature-programmed desorption, K/s

η Site-energy distribution function, J^{-1}

θ_{gr} Gross fractional surface coverage, dimensionless

$\theta(p,q)$ Fraction of sites of energy q covered at pressure p, dimensionless

$\theta(T,E)$ Fraction of sites of energy E covered at temperature T, dimensionless

LITERATURE CITED

1. Hutchinson, H. L., P. L. Barrick and L. F. Brown, Chem. Eng. Prog. Symp. Ser., 63, No. 72, 18 (1967).
2. Adamson, A. W., The Physical Chemistry of Surfaces, 4th ed., pp. 577, 580. Wiley Interscience, New York, 1982.
3. Ross, S., and J. P. Olivier, On Physical Adsorption, esp. pp. 172-3, 189, 249-51. Wiley Interscience, New York, 1964.
4. House, W. A., J. Colloid Interface Sci., 67, 166 (1978).
5. Bräuer, P., W. A. House, and M. Jaroniec, Thin Solid Films, 97, 369 (1982).
6. Phillips, D. L., J. Assoc. Comput. Mach., 9, 84 (1962).
7. Temkin, M., and V. Levich, Zh. Fiz. Khim., 20, 1441 (1946).
8. Sips, R., J. Chem. Phys., 16, 490 (1948).
9. ____, J. Chem. Phys., 18, 1024 (1950).
10. Harris, L. B., Surface Sci., 10, 129 (1968).
11. Drain, L. E., and J. A. Morrison, Trans. Far. Soc., 48, 316 (1952a).
12. Adamson, A. W., and I. Ling, Adv. Chem., 33, 51 (1961).
13. Ross, S., and I. D. Morrison, Surface Sci., 52, 103 (1975).
14. House, W. A., and M. J. Jaycock, Colloid Polym. Sci., 256, 52 (1978).
15. van Dongen, R. H., Surface Sci., 39, 341 (1973).
16. Rudzinski, W., M. Jaroniec, S. Sokolowski and G. F. Cerrofolini, Czech. J. Phys., B-25, 891 (1975).
17. Rudzinsky, W., L. Lajtar and A. Patrykiejew, Surface Sci., 67, 195 (1977).
18. Zolandz, R. R., and A. L. Meyers, Prog. Filt. Sep., 1, 1 (1979).
19. Jaroniec, M., A. Patrykiejew and M. Borowko, Prog. Surf. Membr. Sci., 14, 1 (1981).
20. Jaroniec, M., Adv. Colloid Interface Sci., 18, 149 (1983).
21. King, D. A., Surface Sci., 47, 384 (1975).
22. Tokoro, Y., T. Uchijima and Y. Yoneda, J. Catal., 56, 110 (1979).
23. House, W. A., and M. J. Jaycock, J. Chem. Soc., Faraday Trans. 1, 73, 942 (1977).
24. ____, J. Chem. Soc., Faraday Trans. 1, 74, 1045 (1978)
25. House, W. A., M. Jaroniec, P. Bräuer and P. Fink, Thin Solid Films, 85, 87 (1981).
26. ____, Thin Solid Films, 87, 323 (1982).
27. Papenhuijzen, J., and L. K. Koopal, in Adsorption from Solution (R. H. Ottewill, C. H. Rochester, and A. L. Smith, eds.), pp. 211-225. Academic Press, London, 1983.
28. Butler, J. P., J. A. Reeds and S. V. Dawson, SIAM J. Num. Anal., 18, 381 (1981).
29. Twomey, S., J. Assoc. Comp. Mach., 10, 97 (1963).
30. Tihonov, A. N., Soviet Math. Dok., 4, 2035 (1963a).
31. ____, Soviet Math. Dok., 4, 1624 (1936b).
32. Tikhonov, A. N., and V. Y. Arsenin, Solutions of Ill-Posed Problems, W. H. Winston, New York, 1977.
33. Holt, J. N., and D. L. B. Jupp, J. Inst. Math. Appl., 21, 429 (1978).

34. de Boer, J. H., *The Dynamical Character of Adsorption*, 2nd ed., p. 148. Oxford University Press, London, 1968.

35. Britten, J. A., M.S. Thesis, University of Colorado, Boulder, CO, 1981.

APPENDIX

VARIATIONAL TREATMENT OF EQ. (5) TO OBTAIN EQ. (6)

Starting with Eq. (5), but substituting $[g(t)]^2$ for $H(g)$:

$$I = \int_c^d \left[f_e(x) - \int_a^b K(x,t)g(t)dt \right]^2 dx + \alpha \int_a^b [g(t)]^2 dt \quad . \quad \text{(A-1)}$$

The function $g(t)$ is varied by a small amount $\delta g(t)$ to give

$$I + \delta I = \int_c^d \left[f_e(x) - \int_a^b K(x,t)g(t)dt - \int_a^b K(x,t)\delta g(t)dt \right]^2 dx + \alpha \int_a^b [g(t) + \delta g(t)]^2 dt \quad . \quad \text{(A-2)}$$

Squaring the indicated terms, subtracting Eq. (5), eliminating as negligible terms containing $[\delta g(t)]^2$, and canceling where possible results in

$$\delta I = -2 \int_c^d \left\{ [f_e(x) - \int_a^b K(x,t)g(t)dt] \int_a^b K(x,t)\delta g(t)dt \right\} dx + 2\alpha \int_a^b g(t)\delta g(t)dt \quad . \quad \text{(A-3)}$$

Combining terms yields

$$\delta I = -2 \int_c^d \int_a^b f_e(x)K(x,t)\delta g(t)dtdx + 2 \int_c^d \left[\int_a^b K(x,z)g(z)dz \cdot \int_a^b K(x,t)\delta g(t)dt \right] dx + 2\alpha \int_a^b g(t)\delta g(t)dt \quad . \quad \text{(A-4)}$$

It is assumed that the functions are sufficiently well-behaved that the order of integration may be exchanged. Carrying this out yields

$$\delta I = -2 \int_a^b \left[\int_c^d f_e(x)K(x,t)dx \right] \delta g(t)dt + 2\alpha \int_a^b g(t)\delta g(t)dt \quad \text{(A-5)}$$
$$+ 2 \int_c^d \int_a^b \int_a^b K(x,z)K(x,t)g(z)\delta g(t)dzdtdx.$$

Exchanging the order of integration in the third term on the RHS, then collecting terms gives the final result:

$$\frac{\delta I}{2} = \int_a^b \left[-\int_c^d K(x,t)f_e(x)dx + \alpha g(t) + \int_c^d \int_a^b K(x,z)K(x,t)g(z)dzdx \right] \delta g(t)dt \quad . \quad \text{(A-6)}$$

Since $\delta g(t)$ is completely arbitrary except at the boundaries [where $\delta g(a) = \delta g(b) = 0$], the term inside the square brackets on the RHS must equal zero for a stationary value of I. Setting the term inside the brackets equal to zero yields Eq. (6) in the main text.

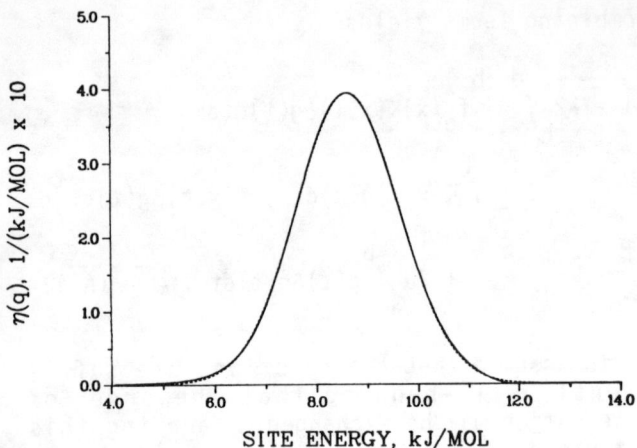

Figure 1. SED by Regularization from Calculated Adsorption Isotherm; Unimodal SED, Langmuir-Isotherm Kernel, Data with 4-Figure Accuracy.
------------ SED postulated to obtain overall adsorption isotherm.
———— SED calculated from isotherm by regularization.

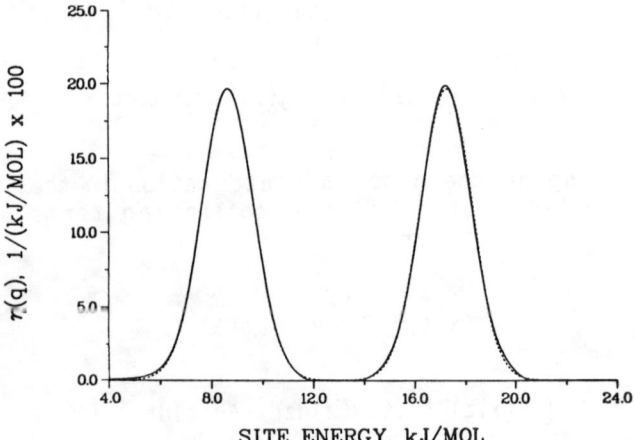

Figure 2. SED by Regularization from Calculated Adorption Isotherm; Bimodal SED, Langmuir-Isotherm Kernel, Data with 4-Figure Accuracy.
------------ SED postulated to obtain overall adsorption isotherm.
———— SED calculated from isotherm by regularization.

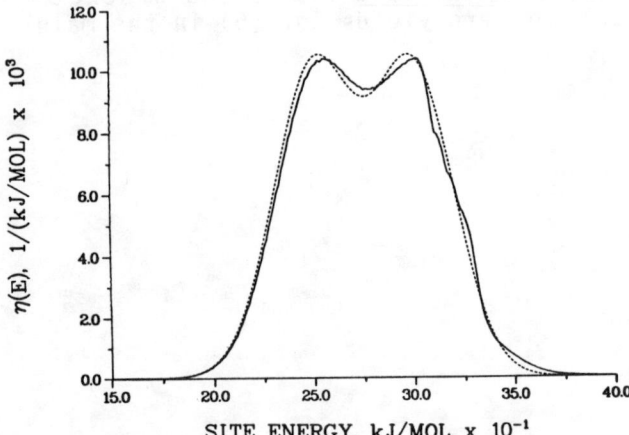

Figure 3. SED by Regularization from Calculated TPD Spectrum; Bimodal SED, First-Order Desorption, Data with 4-Figure Accuracy.
------------ SED postulated to obtain TPD spectrum
———— SED calculated from spectrum by regularization.

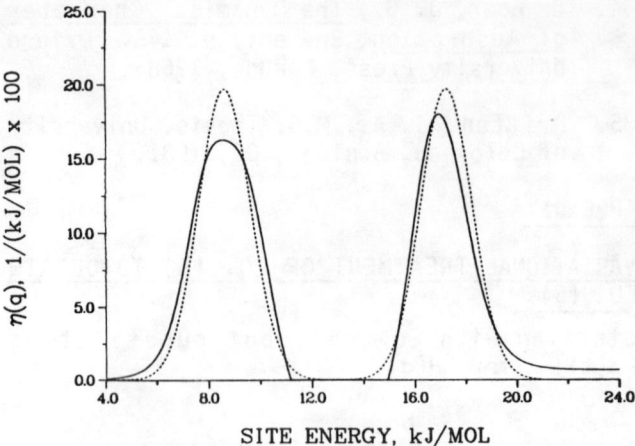

Figure 4. SED by Regularization from Calculated Isotherm with Error; Bimodal SED, Langmuir-Isotherm Kernel, 1% Random Error in Isotherm Data.
------------ SED postulated to obtain overall adsorption isotherm.
———— SED calculated from isotherm by regularization.

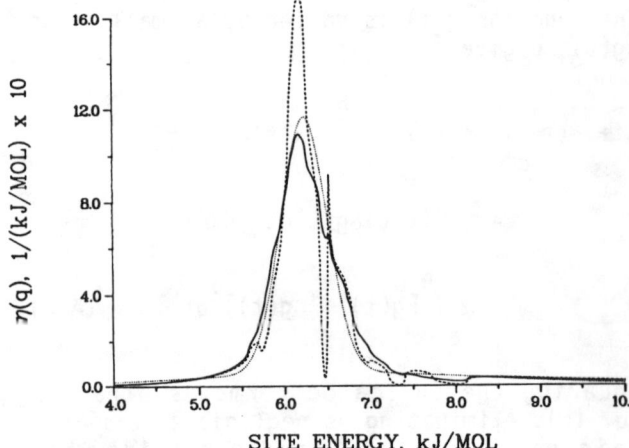

Figure 5. SED by Regularization from Experimental Adsorption Isotherm of Ar on NaBr; Hill-de Boer Isotherm Kernel.
········ SED postulated by Ross and Olivier consistent with data.
------------ SED by regularization at α of BRD criterion.
———— SED by regularization at 0.1α of BRD criterion.

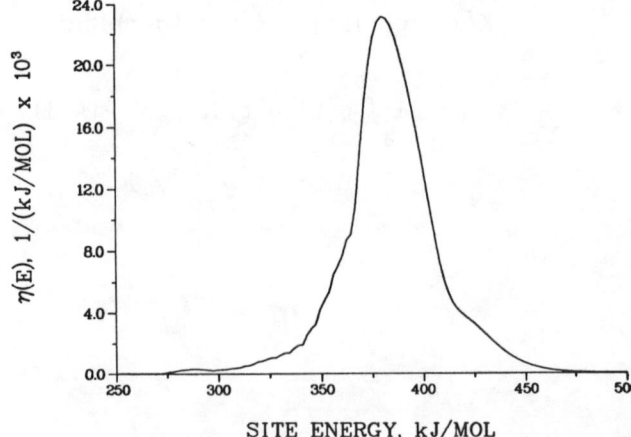

Figure 6. SED by Regularization from Experimental TPD Spectrum of CO from Oxidized Graphite; First-Order Desorption.

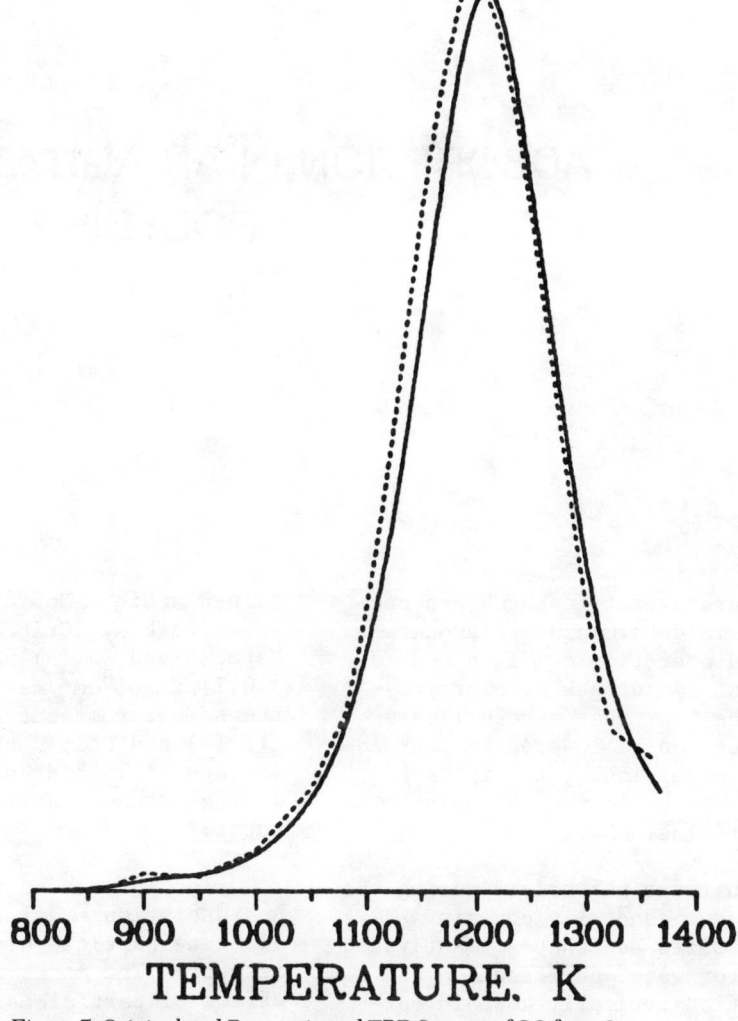

Figure 7. Original and Reconstituted TPD Spectra of CO from Oxidized Graphite.
------------ Experimental TPD Spectrum.
——————— TPD Calculated from SED of Figure 6.

ADSORPTION IN AN AGITATED SLURRY OF POLYDISPERSE PARTICLES

ALEXANDER P. MATHEWS

Department of Civil Engineering
Kansas State University
Manhattan, Kansas 66506

Slurry adsorbers are useful in batch processing of fluids and in determining parameters for the design and operation of fixed bed adsorbers. Important advantages of slurry systems stem from the fact that external mass transfer resistance can be lowered by increasing the agitation power input, and intraparticle resistances can be lowered by using smaller sized particles.

Adsorbents that are pulverized during the manufacturing process generally contain a range of particle sizes and shapes. Activated carbons supplied for water purification and wastewater renovation typically contain particle sizes ranging from Sieve No. 8 to Sieve No. 40. Particle shapes also vary substantially, and this can result in particles with various linear dimensions in a given seive size fraction. The presence of irregularly shaped particles that are extremely long or flat may also result in particles with dimensions larger than the upper sieve size and smaller than the lower sieve size.

Intraparticle diffusivities for spherical particles are expected to be substantially invariant with particle diameter if the pore size distributions do not change significantly for different particle sizes. Mathews and Weber ([1]) obtained a surface diffusivity value of 1.34×10^{-8} cm^2/sec for phenol using 35/40 size (particles passing Sieve No. 35 and retained on Sieve No. 40) Filtrasorb 400 activated carbon. Crittenden and Weber ([2]) and Mathews and Su ([3]) obtained a value of $(3.7 \pm 0.1) \times 10^{-8}$ cm^2/sec for phenol adsorption on the same carbon but of 18/20 size. Fritz et al. ([4]) and Liapis and Rippin ([5]) have found intraparticle diffusivities to vary with the initial concentration of solute in the reactor.

The purpose of this study was to develop an adsorption model that takes into account discrete particle size distributions in the reactor. The model was used to evaluate the effect of particle size distribution within a given size fraction on predicted adsorption rates to explain reported inconsistencies in intraparticle diffusivity values. The model was also used to predict adsorption rates from mixtures of two sieve size fractions and was compared with predictions using a single diameter representing the two sizes.

MATHEMATICAL MODEL

The effect of a Gaussian adsorbent size distribution on adsorption rates in batch reactors has been theoretically studied by Ruthven and Laughlin ([6]) for molecular sieves where the controlling rate mechanism is intraparticle resistance through the micropores. Sensitivity analyses were performed for solutes with linear isotherms, and indicated that both the shape and particle size can significantly affect the adsorption kinetic curves and the diffusivities estimated from these curves and the diffusivities estimated from these curves. Moharir et al. (7) theoretically studied the effect of a particle size distribution on adsorption in fixed beds, again assuming a linear isotherm and intraparticle diffusion-controlled adsorption.

In this work, the homogeneous solid phase diffusion model proposed by Rosen (8) for solutes with linear isotherms was adapted for solutes with nonlinear equilibria, and for a reactor with a range of particle sizes. The model presented here takes into consideration external film transfer of the solute from the liquid phase to the solid surface, equilibrium adsorption at the surface, and subsequent surface diffusion to the interior of the particle.

For the ith particle of radius R_i in a slurry with P particle size fractions, the following rate, equilibrium and material balance equations apply:

External mass transfer is given by

$$\frac{d\bar{q}_i}{dt} = \frac{k_i (C-C_s)}{V_{pi} \rho (1-\varepsilon)} \qquad i = 1,P \qquad (1)$$

where $\bar{q}$ is the average concentration of solute in the particle; k is the external mass transfer coefficient; V_p, ρ and ε are the volume, density and porosity of the particle, respectively; C is the liquid phase solute concentration; C_s is equilibrium liquid phase concentration at the particle surface; and t is time.

Transport within a homogeneous spherical particle assuming symmetry in two directions and a concentration-independent diffusivity is given by

$$\frac{\partial q_i}{\partial t} = D\left(\frac{\partial^2 q_i}{\partial r^2} + \frac{2}{r}\frac{\partial q_i}{\partial r}\right) \qquad i = 1,P \qquad (2)$$

where q_i is the concentration of solute at any point inside the particle; D is the effective surface diffusivity; and r is the radial coordinate.

The average particle concentration is given by

$$\bar{q}_i = \frac{3}{R_i^3} \int_o^{R_i} q_i r^2 dr \qquad i = 1,P \qquad (3)$$

The equilibrium adsorption is represented by the three-parameter isotherm

$$q_s = \frac{aC_s}{(1+bC_s^\beta)}, \beta \leq 1 \qquad (4)$$

where q_s is the equilibrium solute concentration at the particle surface; and a, b and β are isotherm constants.

The material balance for the slurry adsorber is

$$-V\frac{dc}{dt} = \sum W_i \frac{d\bar{q}_i}{dt} \qquad i = 1,P \qquad (5)$$

where V is the reactor volume; and W_i is the weight of adsorbent in each size fraction.

The initial and boundary conditions for $i = 1,P$ are

$$C(0) = C_o$$
$$q_i(r,0) = 0$$
$$q_i(R_i,t) = q_{si}(t) \qquad (6)$$
$$\frac{\partial q_i}{\partial r}(0,t) = 0$$

The solution of equations (1) to (6) was accomplished by a modification of the finite difference procedure reported earlier (Mathews and Weber, 1977).

EXPERIMENTAL METHODS

The adsorbent used in this study was bituminous coal-based activated carbon supplied by the Carborundum Company. The carbon was sieved into four size fractions 12/14, 18/20, 30/35 and 35/40, washed with high purity water and dried at 120°C prior to use. Reagent grade p-chlorophenol was used as the solute.

Adsorption equilibrium studies were conducted by the bottle point method at a constant temperature of 24°C. Powdered carbon of 200 μm to 230 μm size was used for the equilibrium studies. Approximately six days were required for achievement of equilibrium. The solutions were filtered using glass fiber filters (Gelman Science Co.) and analyzed using ultra-violet/visible spectrophotometry. Adsorption rate studies were conducted in a 4.5 liter glass reactor equipped with four baffles. Agitation was provided at constant speed using a four blade stirrer. Adsorption rate studies were conducted at a temperature of 24°C.

PARTICLE SIZE DISTRIBUTION

The size distribution within an individual size fraction was obtained by using a Bausch and Lomb Image Analysis System. The particle dimensions obtained were the projected length and the projected area shown in Figure 1. The diameter based on projected area measurements correspond to the equivalent diameter of a spherical particle with the same area. The mean area per particle was obtained by dividing the total area by the number of particles in the sample. Projected area measurements were made only for the 18/20 and 30/35 sizes. A typical size distribution based on projected length measurement is shown in Figure 2 for the 30/35 sieve size. The particles are well represented by a log-normal distribution for all of the four size fractions analyzed. The log-mean diameter representing 50% of the particles, the length mean, surface mean and Sauter mean (9) are shown in Table 1. In addition, the mean diameters calculated from projected areas for all the particles in the sample are reported in Table 1. For the four size fractions, the geometric means of the sieve openings are found to be substantially lower than all of the other means. The values for log-mean, length mean, surface mean and Sauter mean are close to each other. Hence, for simplicity and comparative analysis only the surface means and the geometric means are considered here.

RESULTS AND DISCUSSION

The adsorption isotherm for p-chlorophenol is shown in Figure 3. The three-parameter isotherm fit to the data points is given by the solid line. The isotherm equation fits the data over a wide concentration range. It is particularly important to have an adequate representation of low and high concentration data since, as shown by Mathews and Su (1981), the adsorption rate profiles are quite sensitive to the isotherm parameters.

Rate data for adsorption of p-chlorophenol on 30/35 carbon at an agitation speed of 600 rpm are shown in Figure 4. The external mass transfer and surface diffusivity coefficients are reported in Table 2. In the case of the geometric mean, a single diameter representing the sieve sizes is used, while in the case of the surface mean, the log-normally distributed particles were represented by three weight fractions, 0.0707, 0.4377, and 0.4916. The diameters corresponding to these weight fractions are 585 µm, 744 µm, and 960 µm, respectively. The estimated coefficients were used to predict adsorption rate profiles for sieve sizes 18/20 and 35/40 at a stirrer speed of 600 rpm and sieve size 12/14 at a stirrer speed of 800 rpm.

Effect of Particle Size on Transport Coefficients

The external mass transfer coefficient and intraparticle diffusivity may vary as a function of particle diameter. There has been very little definitive work on the effect of particle size on external mass transfer coefficients in slurry reactors. Harriot (10) has deduced that for a single particle falling in a stagnant medium at its settling velocity, the mass transfer coefficient should be independent of particle size in the 100 µm to 1000 µm range, should linearly increase with decreasing diameter for particles less than 100 µm, and should vary as $d^{-0.25}$ for particles larger than 1000 µm. The data of Furusawa and Smith (11) for adsorption on activated carbon indicate that the mass transfer coefficient is indeed independent of particle size in the 161-912 µm range used in their study. No data are available for particle sizes outside the 100 µm to 1000 µm range.

In the following analysis, it is assumed that the external mass transfer coefficient is independent of particle size in the 100 µm to 1000 µm range and varies as $d^{-0.25}$ for particles larger than 1000 µm. The mass transfer coefficient for the 30/35 size (543 µm) using the geometric mean was estimated to be 4.18×10^{-3} cm/sec. For the 18/20 and 35/40 size fractions, the same value of 4.18×10^{-3} cm/sec was used in the application of the rate model using geometric mean diameters since these diameters are less than 1000 µm. For the 12/14 size, the mass transfer coefficient was calculated from the equation

$$k_2 = k_1 (d_2/d_1)^{-0.25} \qquad (7)$$

in which k_1 is the mass transfer coefficient at a particle diameter of 1000 µm. The projected lengths of more than 95% of the particles in the 30/35 size are less than 1000 µm. The mass transfer coefficient using the surface mean diameter was estimated to be 8.5×10^{-3} cm/sec. In the application of the rate model using the surface mean diameter, a value of 8.5×10^{-3} cm/sec was used for particles in the 100 µm to 1000 µm range. For particles larger than 1000 µm, the mass transfer coefficient was varied as $d^{-0.25}$.

No attempt was made to measure the pore size distributions for various particle sizes. It was assumed these remain substantially unaltered, and the intraparticle diffusivity is invariant with particle size.

Adsorption rate data for 18/20 carbon at 600 rpm are shown in Figure 5. There is excellent correspondence between experimental and predicted curves using the geometric mean or surface mean. The use of four weight fractions 0.1696, 0.3939 and 0.1426, with the corresponding surface means 999 μm, 1221 μm, 1442 μm, and 1667 μm, respectively, gave better agreement than with the geometric mean. In the case of 35/40 carbon at 600 rpm, shown in Figure 6, the adsorption rate is better predicted with a geometric mean than with the surface mean.

Effect of Agitation Power Input on Mass Transfer Coefficient

The external mass transfer coefficient is correlated well with agitation power input by (12,13).

$$Sh = C_1 \left(\frac{\bar{\varepsilon} d^4}{\nu^3}\right)^{0.25} Sc^{0.33} \quad (8)$$

where Sh and Sc are Sherwood and Schmidt numbers; $\bar{\varepsilon}$ is the power input per unit mass; and ν is the kinematic viscosity. Since $\bar{\varepsilon}$ is proportional to the third power of agitator speed for turbulent flow, it is apparent that the mass transfer coefficient will vary as the 3/4th power of agitator speed as given in equation (9)

$$k_2 = k_1 (n_2/n_1)^{0.75} \quad (9)$$

in which k_1 and k_2 are mass transfer coefficients at agitator speeds of n_1 and n_2 rpm's respectively.

The rate data for the 12/14 particle size at a stirrer speed of 800 rpm are shown in Figure 7. The mass transfer coefficients for both geometric and surface means were adjusted for the effect of particle size and agitator speed using equations (7) and (9). The prediction appears to be excellent when the geometric mean is used, and not as satisfactory with the surface means.

Prediction with Mixed Particle Sizes

The experimental rate data and predictions for a mixture of 18/20 and 30/35 particle sizes at an agitator speed of 600 rpm are shown in Figure 8. The predictions with a single length, surface and Sauter means are compared with the predictions using two geometric means in Table 3. The use of the Sauter mean in a single size fraction model or the use of two geometric mean diameters in a two size fraction model provide better results.

Effect of Particle Diameter and Initial Concentration on Solid Phase Diffusion Coefficients

The results shown in Figures 4 to 8 for five different particle sizes indicate that the adsorption kinetic curve can be predicted using the same intraparticle diffusivity, if the geometric mean diameter is used. Thus, the intraparticle diffusivity is invariant with particle diameter in this system. This is what one would normally expect, if the solute molecular dimensions are small, and the pore size distributions do not appreciably change with particle size diameters. However, for large solute molecules, and adsorbents with pore size distributions that vary with particle size, the results noted above may not apply.

The effect of initial solute concentrations on intraparticle diffusivity values can be examined from Figures 4 to 8. Initial parachlorophenol concentrations ranging from 3.3×10^{-4} M to 5×10^{-4} M (42 mg/ℓ to 64 mg/ℓ) were used in these studies. The data for all these studies are predicted well using a single diffusivity value. Thus, in the concentration range considered, there is no dependence of intraparticle diffusivity on initial solute concentration in the reactor.

CONCLUSIONS

A general adsorption model that takes into account the effects of particle size distributions has been developed. The model provides excellent predictions for mixtures of two particle size fractions when the geometric means for each size fraction are used in the model. A single Sauter mean diameter also provides a good representation of the size distribution.

Adsorption rate studies were conducted with four different adsorbent size fractions. For the parachlorophenol-activated carbon system, the solid phase diffusion coefficient was found to be invariant with the particle diameter when the size fraction was represented by the geometric mean diameter. Also, the diffusion coefficient was found to be invariant with the initial solute concentrations in the reactor, for the concentration range studied.

Particles within a sieve fraction appear to have a range of linear dimensions. Consistently higher values were obtained for mean particle diameters from projected length and projected area measurements compared to the geometric mean of the sieve sizes. The use of projected lengths as the particle diameters in the size distribution is satisfactory for 18/20 and 30/35 sizes. However, for particles larger or smaller than the above sizes, predictions using size distributions were not satisfactory. This may stem from the inadequacy of projected length alone to characterize the particle, and in addition to the projected length, the shape factors may need to be included as a variable. Also, the correlations used to describe the variation of mass transfer coefficient with particle size may not be sufficiently accurate for the larger and smaller sized particles.

ACKNOWLEDGMENT

The assistance of S. R. Kulkarni, graduate student in the Department of Civil Engineering, in the development of the experimental data, and the assistance of Dr. L. A. Glasgow, Assistant Professor in the Department of Chemical Engineering, in obtaining particle size distribution measurements are acknowledged and appreciated.

NOTATION

a, b	three-parameter isotherm constants
a_p	projected area of particle
C	liquid phase solute concentration
C_{ci}	model computed data for liquid phase concentration
C_{ei}	experimental data for liquid phase concentration
C_o	initial liquid phase concentration
C_s	equilibrium liquid phase concentration at particle surface
d	diameter of particle
D	intraparticle diffusivity
i	subscript referring to each size fraction
k	external mass transfer coefficient
P	number of particle size fractions
$\bar{q}$	average particle concentration
q	pointwise concentration inside the particle
q_s	equilibrium solid phase concentration at particle surface
r	radial coordinate
R	particle radius
Sh	Sherwood number
Sc	Schmidt number
t	time
V	reactor volume
V_p	particle volume
W	weight of adsorbent

Greek Symbols

β	isotherm constant
ε	porosity of particle
$\bar{\varepsilon}$	agitation power input per unit mass
ν	kinematic viscosity
ρ	density of particle

LITERATURE CITED

1. Mathews, A. P. and W. J. Weber, Jr., Water-1976: I. Physical, Chemical Wastewater Treatment, AIChE Symp. Ser., 91-97. 73 (166), 91 (1977).

2. Crittenden, J. C. and W. J. Weber, Jr., J. Env. Eng. Div., ASCE, 104, 433 (1978).

3. Mathews, A. P. and C. A. Su, "Modeling Competitive Effects in the Adsorption of Two Organic Priority Pollutants," presented at the annual AIChE meeting, New Orleans (1981).

4. Fritz, W. W., W. Merk, E. U. Schlunder and H. Sontheimer, "Competitive Adsorption of Dissolved Organics on Activated Carbon," in Activated Carbon Adsorption of Organics from the Aqueous Phase, Vol. 1, M. J. McGuire and I. P. Suffet (Eds.), Ann Arbor Science Publishers, Ann Arbor, Michigan (1980).

5. Liapis, A. I. and D. W. T. Rippen, Chem. Eng. Sci., 32, 619 (1977).

6. Ruthven, D. M., and K. F. Loughlin, Chem. Eng. Science, 26, 577 (1971).

7. Moharir, A. S., D. Kunzru, and D. N. Saraf, Chem. Eng. Science, 35, 1795 (1980).

8. Rosen, J. B., J. Phys. Chem., 20, 387 (1952).

9. Foust, A. S., L. A. Wenzel, C. W. Clump, L. Maus, and L. B. Anderson, "Principles of Unit Operations," John Wiley and Sons, Inc., New York (1960).

10. Harriott, P., AIChE J. 8, 93 (1962).

11. Furusawa, T. and J. M. Smith, Ind. & Eng. Chem., 12 197 (1973).

12. Calderbank, P. H. and Moo-Young, M. B. Chem. Eng. Sci. 16, 39 (1961).

13. Brian, P. L. T. and H. B. Hales. AIChE J. 15, 419 (1969).

TABLE 1

MEAN PARTICLE DIAMETERS FOR CARBORUNDUM ACTIVATED CARBON

Sieve Fraction	Geometric Mean (μm)	Log-Mean (μm)	Sauter Mean (μm)	Surface Mean (μm)	Length Mean (μm)	Mean From Projected Area
#12 - #14	1539	1786	1897	1831	1809	--
#18 - #20	917	1200	1282	1231	1215	1238
#30 - #35	543	760	827	786	773	801
#35 - #40	458	560	632	581	566	--

TABLE 2

ESTIMATED MASS TRANSFER AND INTERNAL DIFFUSION COEFFICIENTS FOR PARACHLOROPHENOL

	Diameter (μm)	k (cm/sec)	D (cm^2/sec)
Geometric mean	543	4.81×10^{-3}	1.11×10^{-8}
Surface mean	786	8.5×10^{-3}	1.4×10^{-8}

TABLE 3

MODEL PREDICTIONS FOR PARACHLOROHPENOL FOR MIXTURE OF 18/20 AND 30/35 SIZES

	Diameter (μm)	Percent* Deviation
Length mean	730	10.30
Surface mean	620	8.57
Sauter mean	680	4.34
Two Geometric means	917 543	4.25

* Deviation = $\frac{1}{N} \Sigma \left| (C_{ei} - C_{ci})/C_{ei} \right|$

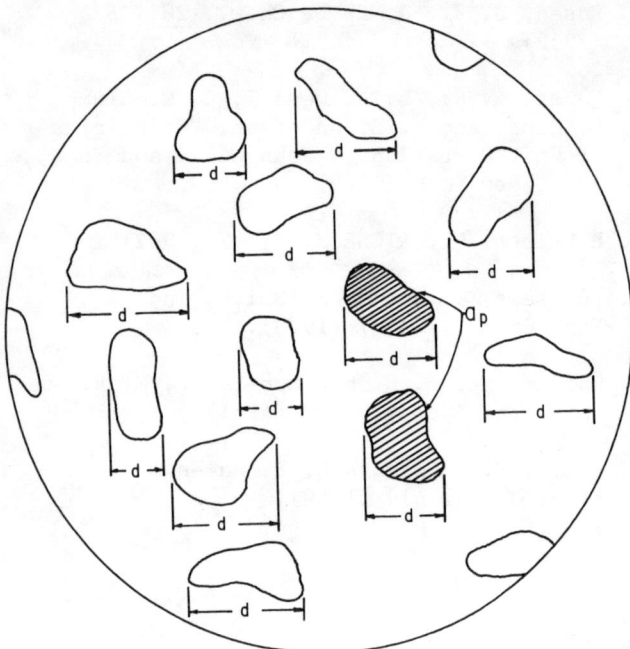

Figure 1. Particle size measurement: projected length (d) and projected area (a_p).

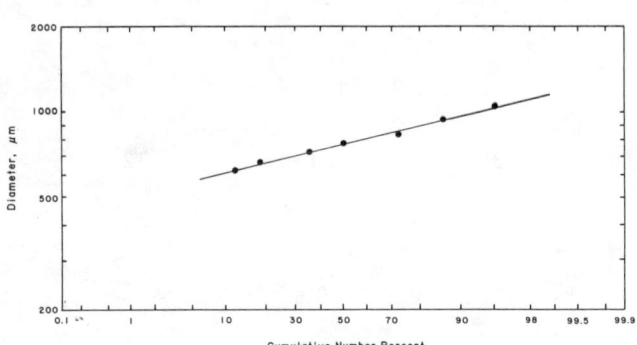

Figure 2. Size distribution for 30/35 mesh size carbon.

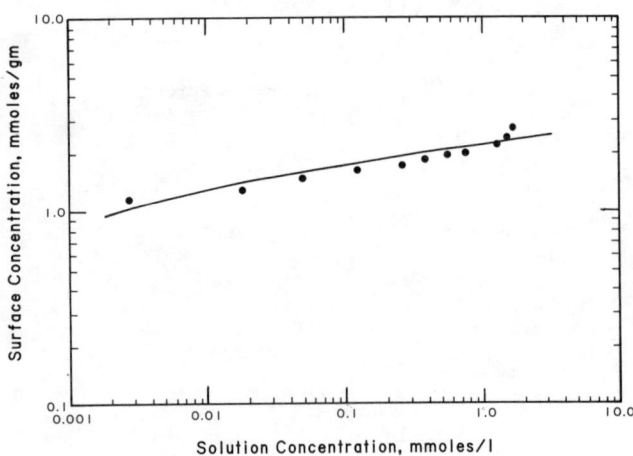

Figure 3. Adsorption isotherm for p-chlorophenol.

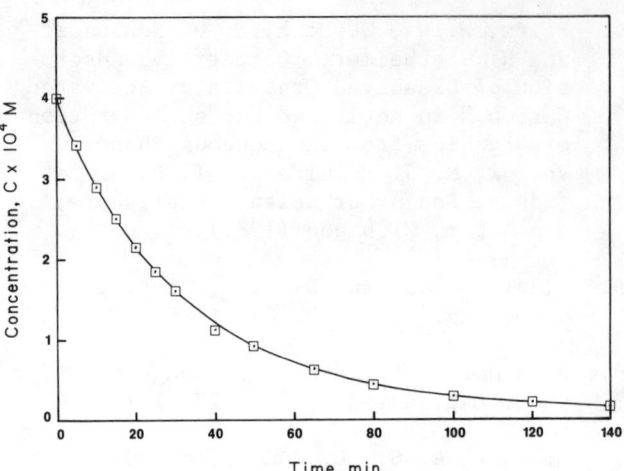

Figure 4. Adsorption rate for p-chlorophenol on 30/35 carbon; (□) experimental data and (−) prediction using geometric mean or surface mean ($C_o = 4 \times 10^{-4}$M, carbon dosage = 0.5 gm/l).

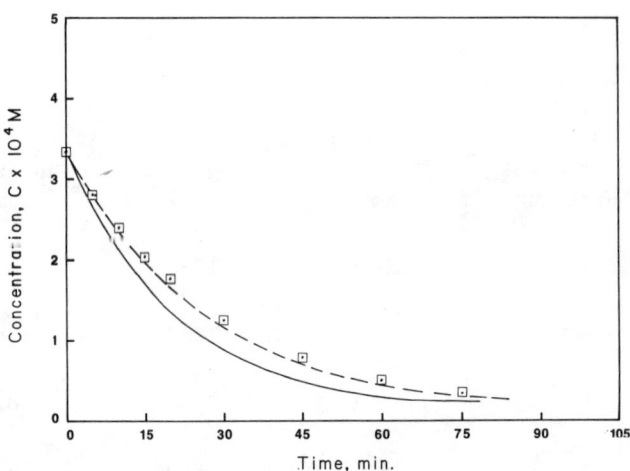

Figure 5. Adsorption rate for p-chlorophenol on 18/20 carbon; (□) experimental data and (−) predictions using (--) geometric mean or surface mean ($C_o = 5 \times 10^{-4}$M, carbon dosage = 0.5 gm/l).

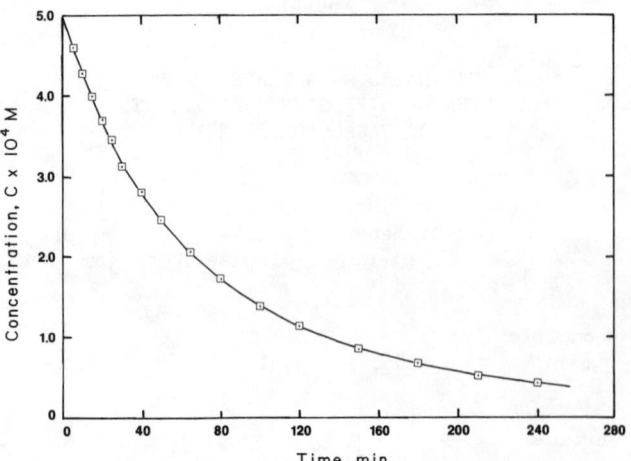

Figure 6. Adsorption rate for p-chlorophenol on 35/40 carbon; (□) experimental data, and predictions using (--) geometric mean or surface mean (−) four surface means ($C_o = 3.35 \times 10^{-4}$M, carbon dosage = 0.67 gm/l).

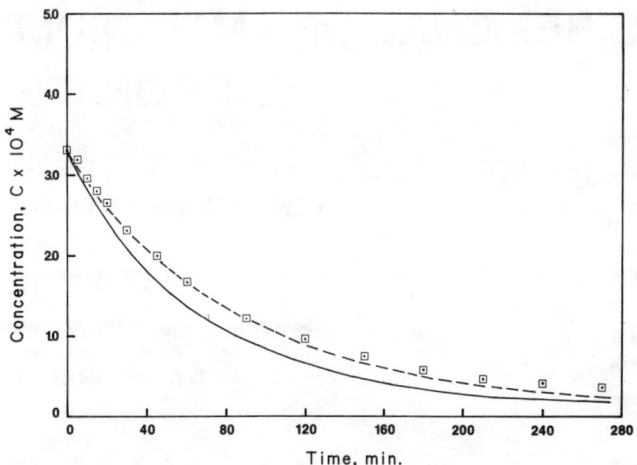

Figure 7. Adsorption rate for p-chlorophenol on 12/14 carbon; (□) experimental data, and predictions using (--) geometric mean and (−) four surface means ($C_o = 3.30 \times 10^{-4}$M, carbon dosage = 0.67 gm/l).

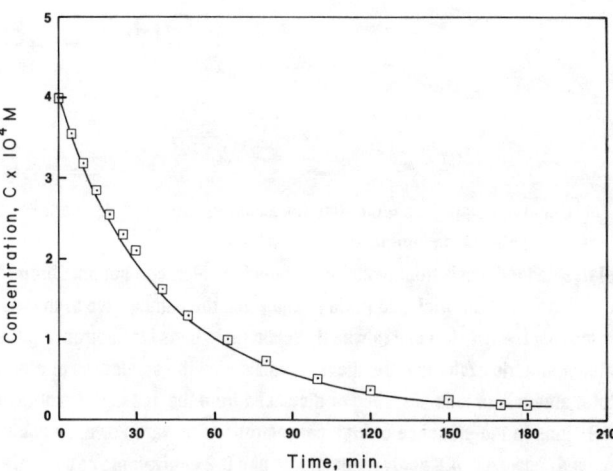

Figure 8. Adsorption rate for p-chlorophenol on 18/20 and 30/35 carbon mixture; (□) experimental data and (−) predictions using two geometric means or single Sauter mean ($C_o = 4 \times 10^{-4}$M, carbon dosage = 0.6 gm/l).

MULTICOMPONENT ION EXCHANGE IN FIXED BEDS FOR SELECTIVE REMOVAL OF AMMONIUM CARBONATE

NIEN-HWA LINDA WANG
and
STANLEY HUANG

School of Chemical Engineering
Purdue University
W. Lafayette, IN 47907

Understanding the competitive ion exchange among NH_4^+, Na^+, K^+, Ca^{++}, and Mg^{++} ions is crucial in designing fixed beds for removing $(NH_4)_2CO_3$ and regenerating the dialysate for dialysis treatments. We applied the Multicomponent Chromatography Theory developed by Helfferich and Klein in analyzing the competitive exchange zeolite beds for removing the NH_4^+ ion and in weakly acidic resin beds for neutralizing the CO_3^{--} ions. For stoichiometric exchange, the theory predicts well the sequences of breakthroughs for the total amounts of ions absorbed or displaced from the beds. The zeolite beds can remove NH_4^+ ions in the presence of high concentrations of Na^+. For each equivalent of NH_4^+ removed, however, 0.8 equivalents of Na^+ and 0.2 equivalents of Ca^{++} are eluted from the zeolite. The resin beds can maintain the effluent pH at about 7 without changing the concentrations of K^+, Ca^{++}, and Mg^{++} in the dialysate. The effluent pH is close to the equilibrium value, which can be controlled by the percent of H^+ loading in the resin. Because the effluent behavior of a single exchanger bed is mainly determined by the separation factors and the interference, the flexibility of controlling the effluent history is limited.

Most of the reported theories and experiments in ion exchange deal with systems in which the competition among different species is unimportant (1). We report here a system for which understanding of the multicomponent interference is crucial. In dialysate regeneration, urea has to be hydrolyzed enzymatically into ammonium carbonate, which then can be removed via ion exchange (2,3). Since dialysate contains a high concentration of physiological salts, selective removal of ammonium carbonate without significantly altering the dialysate composition is a challenging problem. The only dialysate regeneration system commerically available is the Redy system (CCI Life Systems, Oklahoma City, Oklahoma), in which zirconium phosphate preloaded with H^+ and Na^+ is used for removing ammonium carbonate (2-5). It has been reported that zirconium phosphate removes all the K^+, Ca^{++}, and Mg^{++} ions in the dialysate and that the Na^+ concentration continues to rise during dialysis. Nevertheless, a systematic analysis of the competitive ion exchange in the Redy system has never been reported (5).

The purpose of this work is to understand the competitive ion exchange processes and establish a rational design basis for systems in which strict requirement in the effluent history is imperative. In analyzing the competitive processes, we have applied the Multicomponent Chromatography Theory developed by Helfferich and Klein (6,7). The major advantage of this theory is the following. If the separation factors and the total ion exchange capacity of the exchanger are known for given presaturant and influent compositions, one can easily predict the effluent histories and column profiles. The solution involves only finding the roots of polynomials (see next section), whereas in other methods numerical solution of coupled partial differential equations of transport is needed (6,8).

We have tested a zeolite which has a high selectivity for NH_4^+ ions and a weakly acidic resin which can neutralize the CO_3^{--} ions. The total ion exchange capacities for these two exchangers and the separation factors of various ions against Na^+ ions were determined from batch equilibrium experiments. Effluent histories of columns were then determined and compared with the theoretical predictions. We have found that the effluent patterns are well described by the theory. The results also show that the effluent histories of single-exchanger columns cannot meet the requirements for dialysate regeneration, because they are mainly determined from the equilibrium properties and interference. Better control of the effluent histories can be achieved in mixed beds in which two or more exchangers are finely inter-dispersed. The results of mixed beds are given elsewhere (9).

SUMMARY OF THEORY AND PROCEDURE OF CALCULATIONS

The Multicomponent Chromatography Theory was developed by Helfferich and Klein (6). In this theory, one considers how the competition among various ions in a multicomponent system affects the equilibrium distributions of the ions in the mobile and the stationary phases. The major assumptions in the theory are plug flow, stoichiometric ion exchange, constant separation factors, local equilibrium (no mass transfer resistances), and "coherence". Coherence means that all the concentration waves propagating through the column are in phase with each other. Such a state is a stable state and is eventually attainable if there are no disturbances for sufficient time and column length (6).

The assumptions of constant separation factors and coherence lead to the following result. For an N-component system the (N-1) independent concentrations in the mobile phase (x_i's) are related to the separation factors through the following equation (6,9).

$$\sum_i \frac{x_i}{h - \alpha_{1i}} = 0 \qquad (1)$$

The left-hand side of this equation is called the reduced form of the hyperplane (h) transformation, whose general form is

$$H(h, x_i, \alpha_{1i}) = \sum_i \left[\prod_{j \neq i} (h - \alpha_{1j}) x_i \right] \qquad (2)$$

Equation (1) therefore can be expressed in terms of an (N-1) degree polynomial which has real and distinct roots. These (N-1) roots are bounded by the separation factors as follows (6):

$$1 = \alpha_{11} \leq h_1 \leq \alpha_{12} \leq h_2 \leq \alpha_{13} \cdots$$
$$\leq h_{N-1} \leq \alpha_{1N} \qquad (3)$$

For the concentration y_i in the stationary phase, a similar equation and a similar transformation can also be derived (6,9).

$$\sum_i \frac{y_i}{1/h - \alpha_{i1}} = 0 \qquad (4)$$

$$H\left(\frac{1}{h}, y_i, \frac{1}{\alpha_{1i}}\right) = \sum_i \left[\prod_{j \neq i} \left(\frac{1}{h} - \alpha_{j1}\right) y_i \right] \qquad (5)$$

Concentration changes in a coherent system must follow distinct paths, which are called the concentration paths. Along each path only one h-value at a time is allowed to change. Therefore, for an N-component system, there are N possible plateau zones and (N-1) boundaries between these zones. A plateau zone is a finite region of uniform composition in the mobile and stationary phases. By convention, the first plateau zone is the plateau zone near the entrance. A boundary region corresponds to the transition between plateau zones. The boundary is sharp if the composition change is a result of low-affinity species being displaced from the stationary phase by high-affinity species. The boundary is diffuse if the opposite is true. Because the total concentration is fixed, not all the concentrations can increase or decrease simultaneously. If the N components are numbered according to descending order of affinity, there exists an "affinity cut" which divides the N-components into two groups, the high-affinity group and the low-affinity group. The concentrations of the high-affinity group change in the opposite direction as those of the low-affinity group. For example, the first cut is between species 1, which is the most preferred species, and the rest. If the concentration of species 1 decreases from Plateau Zone 1 to Plateau Zone 2, the concentrations of the rest must increase. Only the species marking the affinity cut can be absent from one side of the boundary while being present on the other side of the boundary. All the other species must be either present or absent on both sides of the boundary.

As the boundaries move through the column, the concentrations of the mobile and the stationary phases in a column change with time. Because the mobile phase of the last plateau zone appears first in the effluent, the effluent history corresponds to the reversed column profile. The times at which the boundaries appear in the effluent history can be calculated from the h values and the separation factors.

We now consider the case that a column is uniformly presaturated and the influent composition is abruptly changed to a new value. The procedure for calculating the effluent history is summarized as follows (6,9).

1. Given the presaturant composition x_i'', we find the (N-1) roots of Equation (1), which are called h". We then repeat the

procedure to find h' from the influent composition x_i'. Since all the h values are bounded, they can be found easily by the bisectional method (9).

2. The first plateau zone is in equilibrium with the influent and the last plateau zone is in equilibrium with the presaturant. In order to find the compositions of the plateau zones between the first and the last zones, we change one h' value at a time to h'' to obtain the $(N-2)$ sets of h values for these plateau zones. Only h_k is allowed to change across the kth boundary. We then apply the inverse transformation, Equation (6), to calculate the mobile phase composition of each plateau zone. The mobile phase concentration of species i in Plateau Zone k is given by

$$x_{ik} = \frac{\prod_{j=1}^{k-1}(h_j'' - \alpha_{1i}) \prod_{j=k}^{N-1}(h_j' - \alpha_{1i})}{\prod_{j \neq i}^{N}(\alpha_{1j} - \alpha_{1i})} \quad (6)$$

Once the mobile phase composition in a plateau zone is known, the stationary phase concentration of species i in Plateau Zone k can be easily calculated from

$$y_{ik} = x_{ik} / \sum_j \alpha_{ji} x_{jk} \quad (7)$$

3. If a boundary is sharp, the adjusted velocity of the boundary $\mu_{\Delta k}$ can be calculated from the h values and the separation factors as follows:

$$\mu_{\Delta k} = h_k' h_k'' P_k \quad (8)$$

where $P_k = \prod_{i=1}^{k-1} h_i'' \prod_{i=k+1}^{N-1} h_i' \prod_{i=1}^{N} \alpha_{i1} \quad (9)$

The time scale in the effluent history is expressed in terms of bed volumes of effluent (BV), which is related to the adjusted velocity as follows:

$$BV = \varepsilon + \frac{\bar{c}}{c \mu_\Delta} \quad (10)$$

where ε is the void fraction, $\bar{c}$ is the total capacity per unit bed volume, and c is the total concentration in the mobile phase.

4. If a boundary k is diffuse, we first calculate the adjusted velocity μ_k' of the upstream end and the adjusted velocity μ_k'' of the downstream end from the following two equations:

$$\mu_k' = h_k'^2 P_k \quad (11a)$$

$$\mu_k'' = h_k''^2 P_k \quad (11b)$$

Any intermediate concentration x_i in the diffuse boundary can be calculated from the following equation with a value h_k chosen between h_k' and h_k''.

$$\frac{x_i}{x_{ik}} = \frac{h_k - \alpha_{1i}}{h_k' - \alpha_{1i}} \quad (12)$$

We can again calculate the adjusted velocity corresponding to this x_i value from Equation (8). Finally, we can calculate from Equation (10) the bed volumes of effluent at which the x_i appears in the effluent history. Usually, 10 intermediate x_i's give a good approximation of a diffuse boundary in all the systems tested.

The separation factor of a divalent cation against a monovalent cation is expected to vary with total concentration (6,7,10). We have also found from the results reported below that the variation of separation factor of Ca^{++} against monovalent ions is significant. The change of the separation factors with total concentration can be taken into account by a minor modification of the aforementioned procedure (6, p. 280). The stationary-phase composition y_i'' in equilibrium with the presaturant is calculated from Equation (7) where the mobile phase composition and the separation factors of the presaturant are used. A set of h^* values is then found from Equation (4), where the new separation factors of the influent and y_i'' are used. This set of h^* values instead of those of h'' is used in Equation (6) for calculating the mobile-phase composition. The rest of the treatment remains the same as for the case with constant separation factors.

We have developed a computer program for calculating the effluent history and column profiles for a uniformly presaturated column with a step input. The details of the program are given elsewhere (9).

EXPERIMENTAL

Materials

The zeolite used was Linde Ionsiv IE 95 from Union Carbide. It has a high selectivity for NH_4^+. A weakly acidic resin, Amberlite IRC 50 from Rohm and Haas, was used for supplying H^+ ions to neutralize CO_3^{--} ions. The solution in most experiments was designed to simulate dialysis fluids commonly used in dialysis treatments. This solution is called "dialysate" here. It was normally prepared from chloride salts (AR grade) of the following ions: Na^+(135 mN), Ca^{++}(5.0 mN), K^+ (3.5 mN) and Mg^{++}(2.0 mN). Ammonium carbonate (15 and 50 mN) was added to the dialysate in certain experiments to simulate spent dialysate.

Apparatus and Procedures

Standard procedures for determining the maximum ion exchange capacities and the binary ion exchange isotherms were used (9). All the binary exchange experiments and column experiments were carried out at room temperature (25±3°C). The column experiments were carried out in glass columns with an inner diameter of 1 cm. A peristaltic pump was used in all the column experiments. The flow rate was normally 2.5 ml/min. The effluent from the column was collected in 20 ml portions and analyzed. The concentrations of Na^+ and K^+ were determined from flame emission and those of Ca^{++} and Mg^{++} from atomic absorption. The spectrophotometer was Model 2380 from Perkin Elmer. The concentrations of NH_4^+ ions were measured with an ammonia electrode and a potentiometer (Ion Analyzer, Orion, Model 901). The pH values of the effluents were measured immediately after collection. Details of the column experiments and the analytical procedures have been reported elsewhere (9). The experimental conditions for the column experiments are summarized in Table 1.

RESULTS AND DISCUSSION

Separation Factors and Affinity Sequences

The binary ion-exchange isotherms of the major cations K^+, Ca^{++}, Mg^{++}, and NH_4^+ against Na^+ for the zeolite are shown in Figure 1. They were determined for a total cation concentration of 167 mN. For each binary ion exchange, a separation factor α_{i-Na} was calculated from each data point. An average separation factor for each binary exchange was then estimated from a plot of α versus x_i. The details of the procedure and the estimated uncertainties of these separation factors are given elsewhere (9). The calculated isotherms for the average separation factors are shown as dashed curves in Figure 1. Because $y > x$ for K^+, NH_4^+, and Ca^{++}, these ions are more preferred over Na^+, whereas Mg^{++} ions are less preferred. It can be inferred from Figure 1 that the assumption of constant separation factor is satisfactory for K^+, Ca^{++}, and Mg^{++} but not for NH_4^+. For the latter, large deviations are observed for x larger than 0.3. These deviations are not important for our experiments because the NH_4^+ mole fractions used are smaller than 0.3. The gradual decrease of the separation factor of NH_4^+ against Na^+ is an indication of heterogeneity of the ion exchange sites (11). This is consistent with the manufacturer's information that in this zeolite there are two sizes of pore openings, 2.6 Å and ~ 4 Å (12).

Similar binary exchange isotherms of Ca^{++} and Mg^{++} against Na^+ for a total cation concentration of 167 mN for the resin are shown in Figure 2. The resin clearly has a strong preference for Ca^{++} and Mg^{++}. The separation factors of K^+, Na^+ and NH_4^+ were not determined from the batch experiments. Instead, they were estimated from the effluent history as the best fit parameters for the Multicomponent Theory as described below in the section on effluent histories of the resin beds.

We can convert the separation factors of individual ions against Na^+ to α_{1i}, where species 1 is the most preferred species. These separation factors for the resin and the zeolite are shown in Table 2. These average separation factors and total capacities of 2.3 meq/g for the zeolite and 10.5 meq/g for the resin are employed in calculating the predicted effluent histories for the column experiments.

The separation factors of divalent ions against monovalent ions change with total concentration. For the zeolite and total concentrations from 110 to 170 mN, the variation of α_{13} with total concentration c can be described by the following equation (9).

$$\alpha_{13} = 0.0341 c - 1.3 \qquad (13)$$

This correction gives satisfactory agreement between the effluent histories and the predictions for all the column experiments reported below. The variation of the K^+-Mg^{++} separation factor (α_{15}) is less important because Mg^{++} is a trace component and the zeolite also has a low affinity for Mg^{++}. The correction for α_{15} is important only when the NH_4^+ concentration in the influent is 50 mN or larger. For these cases, the separation factor α_{15} is corrected by keeping the Ca^{++}-Mg^{++} separation factor constant.

Effluent Concentration Histories for Zeolite Beds

The results in Figures 3 through 5 show that the breakthrough sequences of various ions and the total amounts of ions absorbed or eluted can be predicted by the Multicomponent Chromatography Theory. The breakthroughs are not, however, as abrupt as predicted because of mass transfer effects, which are unaccounted for by the theory.

The loading curve for the zeolite bed (Run 1) is shown in Figure 3. The zeolite was presaturated with Na^+. A step input of dialysate was then fed into the bed. As predicted, the breakthrough of the least preferred species Mg^{++} occurred first and that of the most preferred species K^+ occurred last.

The uptakes of Ca^{++} and K^+ as calculated from the breakthrough curves agree to within experimental precision (~10%) with the predicted values. For Mg^{++}, the agreement is poor because the zeolite has a low affinity for Mg^{++}. Only 0.3% of the ion exchange sites are predicted to be occupied by Mg^{++} when the zeolite column is in equilibrium with the dialysate. The rest are occupied by Na^+ (71.5%), K^+ (20.4%), and Ca^{++} (7.9%). The uncertainty of this small value results in apparent violation of mass balance for Mg^{++} as seen in Figure 3.

The effluent histories of zeolite beds for removing NH_4^+ (Runs 2 and 3) are shown in Figures 4 and 5. The predicted effluent history for Run 3 is also given in Table 3. The separation factors used in the calculation are those given in Table 2 except α_{13} and α_{15}. These separation factors were corrected for the concentration effects as described in the previous section (see Equation (13)).

Since there are five species in this system, there are 5 plateau zones and 4 boundaries in the column profile. Plateau Zone 5 appears first in the effluent history, whereas the Plateau Zone 1 appears last (Figure 4). All the boundaries are sharp except Boundary 1, which is diffuse because K^+ in the stationary phase is displaced by less preferred species NH_4^+, Ca^{++}, and Na^+ (see Table 3).

Because Na^+, Ca^{++}, and Mg^{++} are less preferred than NH_4^+, they are eluted before the NH_4^+ breakthrough (Figures 4 and 5). For each equivalent of NH_4^+ removed, about 0.2 equivalents of Ca^{++} and 0.8 equivalents of Na^+ are eluted as calculated from the effluent history of Plateau Zones 3, 4 and 5 (Table 3) (9). Therefore, the use of zeolite for removing NH_4^+ inevitably results in return of Ca^{++} and Na^+ to the effluent. For this reason, the zeolite is not suitable for dialysate regeneration.

Effluent Concentration Histories for Resin Beds without H^+ Exchange

When there is no H^+ exchange, the total concentration of cations remains constant. The dialysate loading curve of a resin bed preloaded with 45% H^+ and 55% Na^+ is shown in Figure 6. As predicted, the breakthroughs of K^+ and NH_4^+, which are less preferred ions, occurred first and those of Mg^{++} and Ca^{++}, which are the most preferred ions, occurred last. The Mg^{++} peak is a result of strong interference between Mg^{++} and Ca^{++}. The location of the Mg^{++} peak and the absence of Ca^{++} peak are correctly predicted by the theory.

The resin used has very low affinities for NH_4^+ and K^+. The separation factors of Ca^{++}-NH_4^+ and Ca^{++}-K^+ were not determined experimentally. The best estimates based on the column experiments are about 40 and 70. These estimates were used in the theory to determine the predicted effluent curves shown in Figure 6. The uptakes of Ca^{++} and Mg^{++} estimated from the data agree to within the experimental precision with the predicted values.

Because the resin has extremely high affinities for Ca^{++} and Mg^{++}, a large volume of dialysate (167 BV) is needed to saturate the Na^+ sites of the resin (Figure 6). At equilibrium, 57% of the sites in Run 5 are occupied by Ca^{++}, 16% by Mg^{++}, 22% by Na^+, 4% by NH_4^+, and 0.6% by K^+. Such a presaturated resin can act as a reservoir of Ca^{++} and Mg^{++}. As shown in Figure 7, the presaturated resin can supply the mobile phase with Ca^{++} and Mg^{++} when the influent does not contain these ions (Run 6). Eventually, the preloaded Ca^{++} and Mg^{++} will be depleted and the resin will lose its buffering power. Such behavior is correctly predicted from the theory. The data show an earlier depletion than predicted because of mass transfer effects. The depletion is predicted to occur gradually, because it corresponds to a diffuse boundary between two plateau zones. The boundary is diffuse because the preloaded Ca^{++} and Mg^{++} are displaced by Na^+, which is a less preferred species.

Effluent Concentration Histories for Resin Beds with H^+ Exchange

For a weakly acidic resin, the release of H^+ ions is controlled by the ionization constant of the functional groups and the pH of the solution (13). When an influent contains CO_3^{--} ions, the influent pH is high and the H^+ ions are released to neutralize the CO_3^{--} ions. The pH of the solution in equilibrium with the resin is controlled by the fraction of H^+ loading as indicated by the titration curve of the resin (9). The effluent pH's for two resin beds with 15% (Run 6) and 45% (Run 7) H^+ preloadings are shown in Figure 8. The effluent pH varies from 7.0 to 7.5 for the former and from 6.5 to 7.0 for the latter. These values are in agreement with the equilibrium pH of the titration curve (9).

The effluent Ca^{++} and Mg^{++} histories for Runs 6 and 7 are shown in Figures 9 and 10. These effluent histories are similar to those of Run 5, whose influent contains no Ca^{++} and Mg^{++}. The reason is the following. As H^+ ions are released to neutralize the CO_3^{--} ions, the sites which are previously occupied by H^+ remove all the Ca^{++} and Mg^{++} from the mobile phase. Since there are not enough Ca^{++} and Mg^{++} ions in the mobile phase, these sites are temporarily occupied by Na^+. The Na^+ ions in these sites continue to exchange with the Ca^{++} and Mg^{++} ions in the mobile phase. As a result, the portion of the resin bed upstream from the H^+ front acts as a sink for Ca^{++} and Mg^{++}, whereas the portion downstream from the H^+ front acts as a Ca^{++} and Mg^{++} reservoir. Consequently, the effluent Ca^{++} and Mg^{++} histories of Runs 6 and 7 are similar to those of Run 5 except that for the former runs the reservoirs shrink as the H^+ fronts advance. We have developed a shrinking reservoir model which predicts well the effluent histories of Ca^{++} and Mg^{++} for systems with H^+ exchange. The detailed calculations are reported elsewhere (9).

From this model, the Ca^{++} and Mg^{++} plateau zones end at 18 bed volumes of effluent for Run 6 (Figure 9) and at 15 bed volumes for Run 7 (Figure 10). The speed of H^+ front should be three times smaller for Run 7 (45% H^+ preloading) than for Run 6 (15% H^+ preloading). This implies that a larger portion of the bed in Run 7 acts as a Ca^{++} and Mg^{++} reservoir. Nevertheless, the amounts of Ca^{++} and Mg^{++} preloaded per unit bed volume for Run 7 are about 65% of those for Run 6. These two effects almost counterbalance each other. Overall, the plateau zones of Ca^{++} and Mg^{++} for these two runs are similar.

Remarks on the Use of Fixed Beds for Dialysate Regeneration

In dialysate regeneration, one wishes to (i) remove NH_4^+ ions, (ii) neutralize CO_3^{--} ions, (iii) control the effluent pH at about 7.4, and (iv) maintain Na^+, K^+, Ca^{++}, and Mg^{++} concentrations constant to within the physiological limits. As discussed above, a two layer bed of the zeolite and the resin can achieve the first three goals. It can not, however, achieve the last goal. For each equivalent of NH_4^+ removed, 0.8 equivalents of Na^+ (Figure 5b) and 0.2 equivalents of Ca^{++} (Figure 5a) will be added to the dialysate before the NH_4^+ breakthrough.

In general, the effluent histories of single exchanger columns are mainly controlled by the equilibrium properties and the interference of the various ions. Although the effluent history can be adjusted to some extent by varying the presaturation conditions, there is a limited flexibility. A more effective way of controlling the effluent history of a column is to use mixed beds which consist of two or more exchangers finely interdispersed. Such mixed beds can have an effluent history substantially different from those of single exchanger beds. The effluent histories of mixed beds can be changed over considerable ranges by choosing different combinations and by adjusting the relative amounts of the individual exchangers. We have designed mixed bed systems in which ammonium carbonate is removed whereas the amount of Na^+ added to the fluid is reduced by the proper H^+ loading and proper bed composition. The effluent history has no Ca^{++} eluted before the NH_4^+ breakthrough. These results are given elsewhere (9). The Multicomponent Chromatography Theory is useful in understanding the competitive exchange in mixed beds and in determining the design parameters such as the choice of exchangers, their relative amounts, and the preloading conditions.

CONCLUSIONS

1. For stoichiometric exchange in the zeolite and resin beds, the Multicomponent Chromatography Theory of Helfferich and Klein predicts well the effluent sequences and the total amounts of ions absorbed or eluted. Because of mass transfer effects, the breakthrough curves are not as abrupt as predicted. The separation factors for Na^+, K^+, and NH_4^+ can be taken to be constant in all the systems tested. In contrast, the separation factors for Ca^{++} and Mg^{++} change significantly with total concentration.

2. The major characteristics of the effluent history, namely the sequence of breakthroughs, elution peaks, and uptakes, are controlled mainly by the separation factors and the interference among the various ions. For single exchanger beds, the flexibility of controlling the effluent history by preloading conditions is limited.

3. The zeolite beds tested can remove NH_4^+ ions in the presence of a high concentration of Na^+. For each equivalents of NH_4^+ removed, however, 0.8 equivalents of Na^+ and 0.2 equivalents of Ca^{++} are returned. For this reason, the zeolite tested can not meet the needs for dialysis treatments.

4. The resin beds tested can neutralize the CO_3^{--} ions and maintain the effluent pH at about 7. The effluent pH is close to the equilibrium value, which can be controlled by the percent of H^+ preloading on the resin.

ACKNOWLEDGEMENT

This research is partly supported by NSF grant CPE 8105959.

LITERATURE CITED

1. Vermeulen, T., G. Klein, and N.K. Heister, "Adsorption and Ion Exchange," in Chemical Engineers' Handbook, R.H. Perry and C.H. Chilton (eds.), 5th ed., McGraw-Hill, New York (1973).

2. Blumenkrantz, M.J., A. Gordon, M. Roberts, A.J. Lewin, E.A. Pecker, J.K. Moran, J.W. Coburn, and M.H. Maxwell, Artificial Organs, 3 (3), 230 (1979).

3. Gordon, A., O.S. Better, M.A. Greenbaum, L.B. Marantz, T. Gral, and M.H. Maxwell, Trans. Amer. Soc. Artif. Int. Organs, 17, 253 (1971).

4. Gordon, A., Lewin, A., Marantz, L.B., and Maxwell, M.H., Kidney International, 10 (Suppl. 7), S 277 (1976).

5. Henderson, L.W., ASAIO Journal, 1 (2), 49 (1979).

6. Helfferich, F. and G. Klein, Multicomponent Chromatography; Theory of Interference, Dekker, New York (1970).

7. Klein, G., D. Tondeur, and T. Vermeulen, Ind. Eng. Chem. Fundam., 6 (3), 339 (1967).

8. Rhee, H.K., R. Aris, and N.R. Amundson, Phil. Trans. Roy. Soc. Lond. A., 267 (1182), 419 (1970).

9. Huang, S., "Multicomponent Ion Exchange in Mixed Beds for Dialysate Regeneration," M.S. Thesis, Purdue University (1983).

10. Helfferich, F., Ion Exchange, McGraw-Hill, New York (1958).

11. Amphlett, C.B., Inorganic Ion Exchangers, Elsevier, New York (1964).

12. Ion Exchange Bulletin, Linde Molecular Sieves, Union Carbide, New York.

13. Klein, G., J. Sinkovic, T. Vermeulen, L.H. Williams, and J. Dong, "Weak-Electrolyte Ion Exchange in Advanced-Technology Water-Reuse Systems," OWRT report, U.S. Dept. Interior, OWRT/RU-82/7 (1982).

Table 1. Experimental Conditions for the Column Experiments

RUN#	EXCHANGER Zeolite Vol., ml	EXCHANGER Resin Vol., ml	EXCHANGER %H+ loading	PRESATURANT	INFLUENT [NH_4^+]	INFLUENT [CO_3^{-2}]	INFLUENT
1	11.6	0		135 mN Na+	0	0	dialysate*
2	11.6	0		dialysate	15	0	dialysate
3	8.9	0		dialysate but 100 mN Na+	50	0	dialysate
4	0	7.5	45	150 mN Na+	15	0	dialysate but 150 mN Na+
5	0	7.5	45	dialysate but 150 mN Na+	0	0	150 mN Na+
6	0	10.8	15	dialysate but 150 mN Na+	0	15	dialysate but 150 mN Na+
7	0	7.5	45	dialysate but 150 mN Na+	0	15	dialysate but 150 mN Na+

*Composition given in Materials Section.

Table 2. Cation Affinity Sequences and Average Separation Factors for the Zeolite and the Resin+

Zeolite

Affinity sequence	1	2	3	4	5
Species	K^+	NH_4^+	Ca^{++}	Na^+	Mg^{++}
α_{1i}	1.0	2.2	4.4	11	45.5

Resin

Affinity sequence	1	2	3	4	5
Species	Ca^{++}	Mg^{++}	NH_4^+	K^+	Na^+
α_{1i}	1.0	1.42	40*	70*	78.2

+Total concentration of the cations is 167 mN for all the experiments.

*Inferred from the effluent histories of column experiments and the Multicomponent Theory.

Table 3. Predicted Effluent History of Run 3

Plateau Zone	5	4	3	2	1
Dimensionless Concentrations					
K^+	1.39	1.41	1.74	4.21	1.00
NH_4^+	0.00	0.00	0.00	0.82	1.00
Ca^{++}	2.34	2.37	6.19	0.94	1.00
Na^+	1.39	1.42	1.21	0.98	1.00
Mg^{++}	2.34	1.06	1.04	1.00	1.00
Boundary		4	3	2	1
B.V. of Effluent		2	15	18	45-32
Type		sharp	sharp	sharp	diffuse

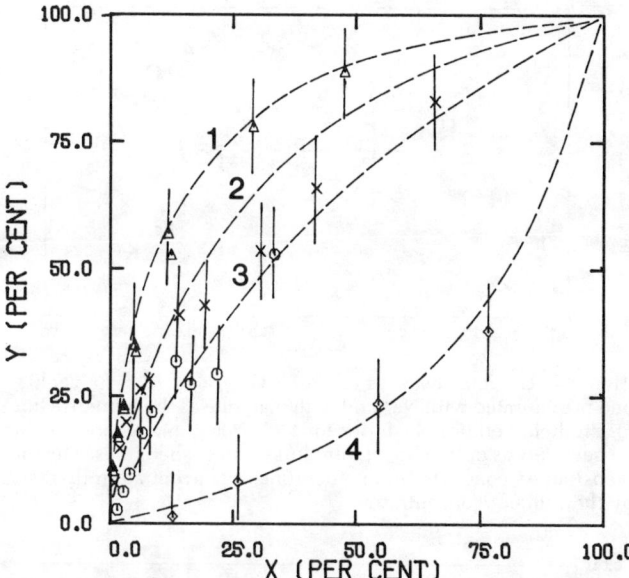

Figure 1. Binary Ion-Exchange Isotherms of $K^+ - Na^+$, ($\triangle$), $NH_4^+ - Na^+$ (x), $Ca^{++} - Na^+$ ($\bigcirc$), and $Mg^{++} - Na^+$ ($\Diamond$) for the zeolite. The dashed lines are calculated isotherms of constant separation factors of 11, 4.6, 2.5, and 0.22.

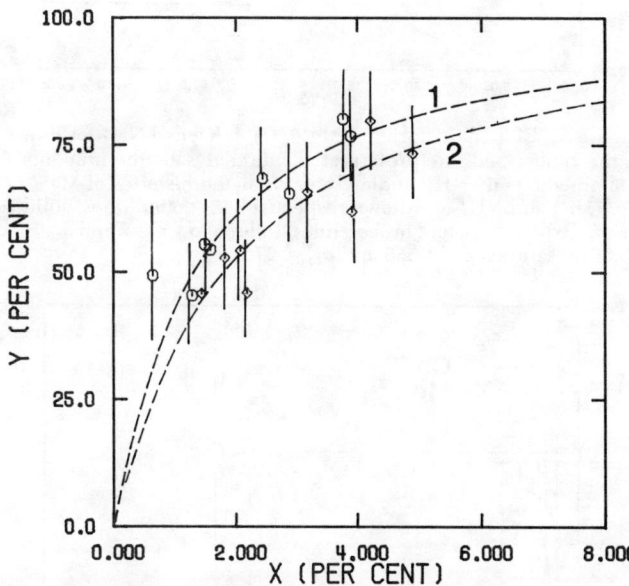

Figure 2. Binary Ion-Exchange isotherms of $Ca^{++} - Na^+$ ($\bigcirc$) and $Mg^{++} - Na^+$ ($\Diamond$) for the resin. The dashed lines 1 and 2 are calculated isotherms of constant separation factors of 78.2 and 55.0.

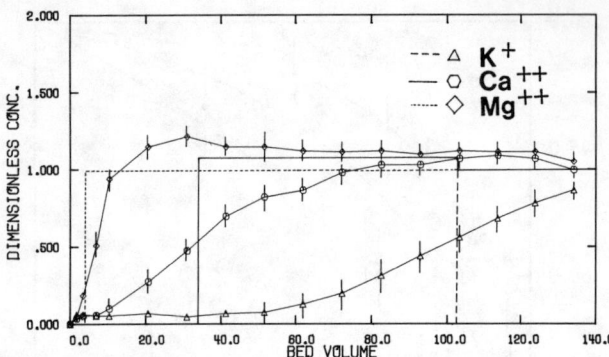

Figure 3. Effluent histories of Mg^{++}, Ca^{++}, and K^+ for a zeolite bed presaturated with Na^+ and with dialysate as the influent (Run 1). Predicted effluent histories for Mg^{++}, Ca^{++} and K^+ are shown respectively as dotted lines, solid lines, and dashed lines. The dimensionless concentration is the effluent concentration divided by the influent concentration.

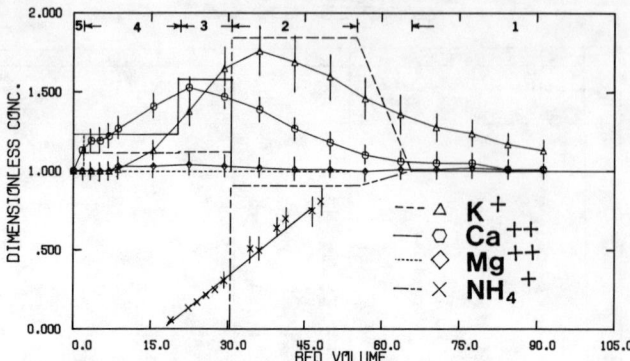

Figure 4. Effluent histories of Ca^{++}, K^+, and NH_4^+ for Run 2; the zeolite bed was presaturated with dialysate; the influent contained 15 mN NH_4^+ in dialysate. Predicted histories for Ca^{++}, K^+, and NH_4^+ are shown respectively as solid lines, dashed lines, and broken lines. Influent $\alpha_{13} = 4.2$; presaturant $\alpha_{13} = 3.7$

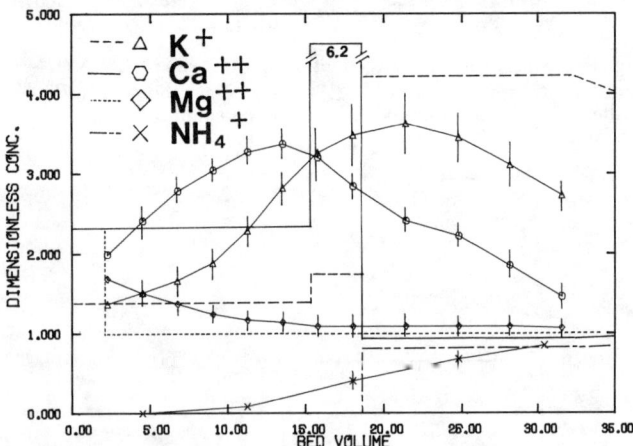

Figure 5a. Effluent histories of Mg^{++}, Ca^{++}, K^+, and NH_4^+ for Run 3; the zeolite beds are presaturated with dialysate; the influents contained 50 mN NH_4^+ in dialysate. Predicted histories of Mg^{++}, Ca^{++}, K^+, and NH_4^+ are shown respectively as dotted lines, solid lines, dashed lines, and broken lines. Influent $\alpha_{13} = 4.2$ and $\alpha_{15} = 45.5$; presaturant $\alpha_{13} = 2.5$ and $\alpha_{15} = 27.1$.

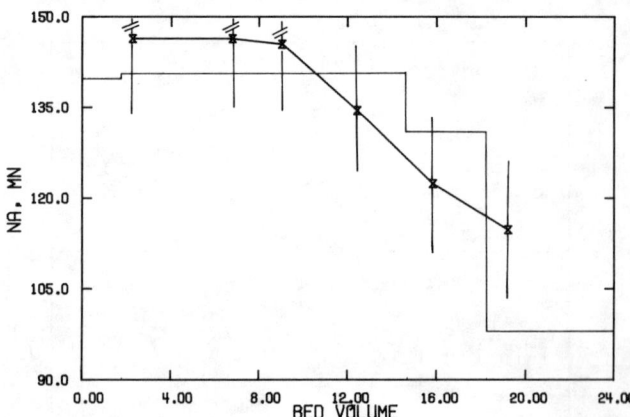

Figure 5b. Effluent histories of Na^+ for Run 3. Predicted Na^+ history is shown as solid lines.

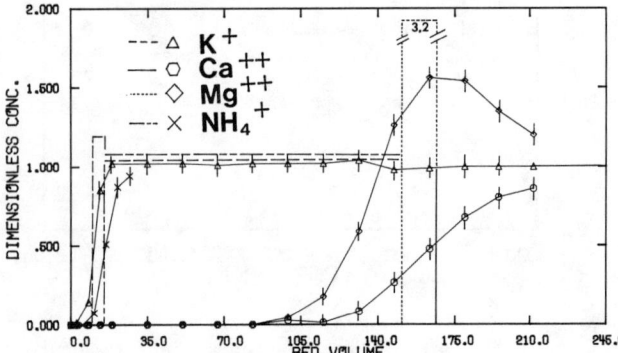

Figure 6. Effluent histories of K^+, NH_4^+, Mg^{++}, Ca^{++} for a resin bed presaturated with Na^+, the influent contained 15 mN NH_4^+ in dialysate (Run 4). Predicted effluent histories of K^+, NH_4^+ Mg^{++}, and Ca^{++} are shown respectively as dashed lines, broken lines, dotted lines, and solid lines.

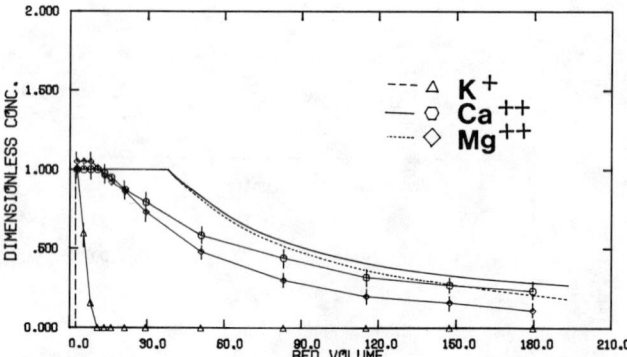

Figure 7. Effluent histories of K^+, Ca^{++}, and Mg^{++} for a resin bed (45% H^+ loaded) presaturated with dialysate and with 150 mN Na^+ as the influent (Run 5). The dimensionless concentrations are based on normal dialysate values. Predicted effluent histories of K^+, Ca^{++}, and Mg^{++} are shown respectively as dashed, solid, and dotted lines.

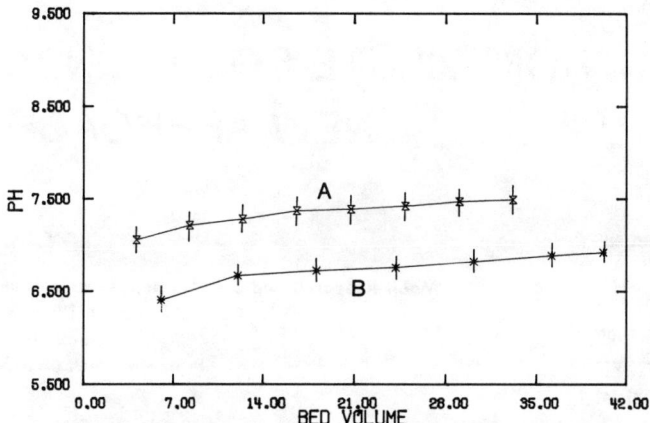

Figure 8. Effluent pH histories for resin beds with 15% (Run 6, Curve A) and 45% (Run 7, Curve B) H$^+$ preloading and with influents of 15 mN CO$_3^{--}$ in dialysate.

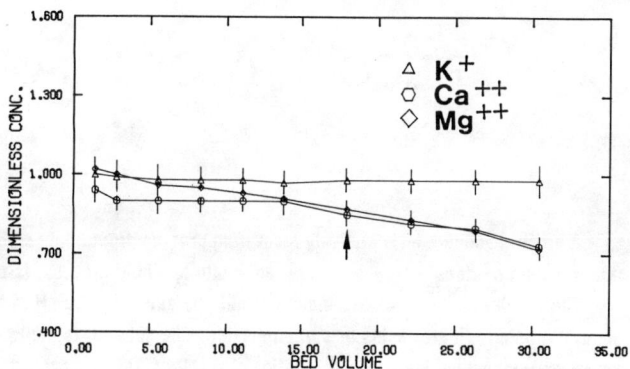

Figure 9. Effluent histories of K$^+$, Ca^{++}, and Mg^{++} for a resin bed (15% H$^+$ preloading) presaturated with dialysate and with an influent of 15 mN CO$_3^{--}$ in dialysate (Run 6). The arrow indicates the end of Ca^{++} and Mg^{++} plateau zones predicted from the shrinking reservoir model.

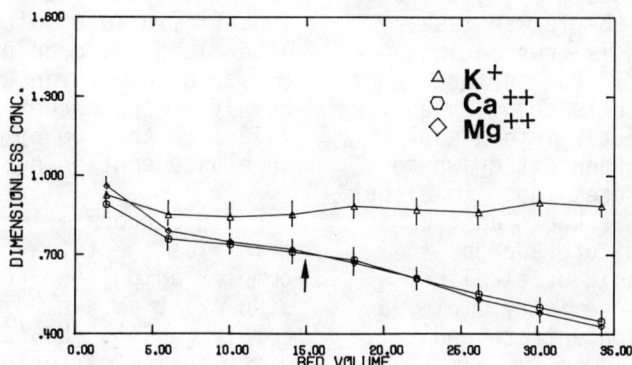

Figure 10. Same as Figure 9 but for 45% H$^+$ preloading (Run 7).

RECOVERY OF URANIUM FROM DILUTE SOLUTION: A NEW APPROACH

A process scheme for effectively recovering uranium from dilute solutions, particularly alkaline carbonate leachates from an in situ leaching process, is presented.

The uranium complexes are made unstable by adjusting the pH of the leachate to about 6.5 using mineral acids or carbon dioxide. The solution is then passed over ion exchange resin which induces precipitation of uranium. This nonexchangeable uranium on the resin is then eluted or "leached" with acid or carbonate solution to obtain eluate of high uranium concentration.

The scheme provides a technique to load the uranium onto a resin beyond its ion exchange capacity resulting in loading capacities 2 to 3 fold greater than that of conventional ion exchange processes.

The scheme is particularly effective for recovering uranium from solutions containing high levels of competing anions, such as Cl^- and SO_4^{2-}.

T. Y. YAN

Mobil Research and Development Corporation
Central Research Division
P.O. Box 1025
Princeton, N.J. 08540

Uranium is generally recovered from it's ores by leaching. The leaching may be carried out in conjunction with a surface milling operation or performed in situ. In the former case, uranium obtained by mining is crushed and blended prior to leaching in the surface facilities. During in situ leaching, the lixiviant is injected directly into a subterranean ore deposit and then withdrawn to the surface for further processing. In either case the leachate, sometimes known as the pregnant lixiviant, is a dilute aqueous uranium solution, either acid or alkaline, depending on the leaching chemistry employed, which must be treated to concentrate and recover the uranium.

Ion exchange and solvent extraction are the most commonly used techniques for recovering uranium from acidic leachates. However, for alkaline carbonate leachates, solvent extraction has not proved economically feasible, so that ion exchange is the standard practice (1). Uranium recovery from the leachate is an important and integral part of the in situ leaching process and improvements in the ion exchange process used can contribute significantly to improving the overall process. In this paper a new approach to uranium recovery from dilute solution is described. Recovery of uranium from carbonate leachates, particularly from in situ leaching operations, is stressed even though the technique is more general in its applications.

RESIN ION EXCHANGE

The ion exchange resin adsorbs uranium ions from dilute solution in a loading cycle. After loading, uranium is eluted to obtain an eluate of high uranium concentration. The net result is to concentrate the uranium 50 to 100 fold from the 100 ppm typical of a leachate to an eluate containing about 5 to 10 g/ℓ.

In carbonate leachates, the uranium exists mainly as the tetravalent uranyl tricarbonate complex anion, $[UO_2(CO_3)_3]^{4-}$, although it can also exist as the divalent uranyl dicarbonate complex anion, $[UO_2(CO_3)_2]^{2-}$. The equilibrium distribution of these two forms depends on carbonate concentration, and particularly, on the pH of the solution. The equilibrium constant has been reported by Blake et al. (2) and Garrels and Hostetler (3), and concentrations of the complexes have been determined by Lemoine (4) and Naumov (5).

Suitable resins for adsorbing uranyl carbonate complexes are anionic, particularly strong base anionic resins containing quaternary ammonium functional groups as the active ion constituent. Typical chemical structures are:

Type I

$$\left[\begin{array}{c}\text{-CHCH}_2\text{-}\\ \bigcirc\text{---CH}_2\text{N(CH}_3)_2\\ |\\ \text{C}_2\text{H}_4\text{OH}\end{array}\right]^{\oplus} \text{Cl}^-$$

Type II

The typical reactions involved in the loading cycle are:

$$4RX + [UO_2(CO_3)_3]^{4-} \rightleftarrows R_4UO_2(CO_3)_3 + 4X^- \quad (1)$$

$$2RX + [UO_2(CO_3)_2]^{2-} \rightleftarrows R_2UO_2(CO_3)_2 + 2X^- \quad (2)$$

The reactions involved in the elution cycle are just the reverse of the above.

To improve the efficiency of ion exchange operation, the loading capacity of the resin should be maximized. The loading capacity of a resin is most importantly determined by two factors, namely, the total ion exchange capacity and the selectivity for uranium. The total ion exchange capacity is unique to the chemical structure and rather difficult to increase. A survey of the commercially available strong base anion resins showed that the total ion exchange capacity varies within a narrow range of 1.0 to 1.4 meq/cc.

Since the leachate contains many competing ions such as Cl^-, SO_4^{2-} and HCO_3^- in high concentrations, a high selectivity for uranium would result in higher uranium loading capacity. The selectivity is also greatly determined by the chemical structure of the resin. For example, type II resins have been found to be more selective for uranium in the chloride solution than the type I resins (6).

This investigation, however, concerns another avenue to increase uranium loading capacity, by changing the nature of the uranium complex itself.

It is apparent from Equation (1) and (2) that, if the pH and carbonate concentration of the solution is lowered to make the uranyl dicarbonate (UDC) the predominant species, the loading capacity will have doubled. However, the loading capacity is still subject to the constraint of the total ion exchange capacity of the resin. In addition, the selectivity of the resin for UDC is significantly lower than the uranyl tricarbonate (UTC), so that the uranium loading capacity cannot double in the presence of other competing anions. We have found, however, that by properly controlling the loading conditions, it is possible to load the uranium in the form of a non-exchangeable precipitate, and hence to deposit quantities of uranium much beyond the theoretical ion exchange capacity. In addition to being free from the constraint of ion exchange capacity, this approach is also free from the detrimental effects of competing ions.

The non-exchangeable uranium on the resin is recovered by leaching with carbonate, preferably acid solution in lieu of conventional eluant (7).

EXPERIMENTAL

Resins

Commercial resins A, B and C were used as received. Their chemical and physical properties are shown in Table 1.

Feed Solutions

The feed solutions were prepared to simulate leachate by dissolving reagent grade chemicals in distilled water. The pH of the final solutions were adjusted with H_2SO_4 to the desired values. The compositions of the feed solutions are shown in Table 2.

Eluant

1 N HCl and/or 1 N NaCl solution were used.

Uranium Loading

3 cc ($\sim$ 2 g) of resin was packed in a glass tube of 0.4 cm i.d., resulting in a bed height of 25 cm. The feed solution upflowed through the bed at 20 cc/hr rate (0.65 gpm/ft^2). The effluent was analyzed for uranium using the modified Bromo-PADAP colorimetric method (8). Based on the uranium concentration analyses of the effluents, the loading curve was constructed and the loading capacities at various levels of uranium leakage were calculated. Comparison of several loading cycles suggest that these loading curves are repeatable within ± 5%. The uranium loading capacity calculated based on the loading curve tends to be conservative and somewhat lower than that based on the elution result. The uranium loading capacity is expressed as lb of U_3O_8/ft^3 of resin. For the purpose of comparing different resins the loading capacities at 50 ppm (25%) uranium leakage were calculated and tabulated.

Elution

The resins loaded with uranium were eluted with 1 N HCl at 10 cc/hr rate. The eluate was analyzed for uranium, and the elution curves were constructed. The total uranium eluted was calculated and compared with that from the loading cycles to obtain uranium material balance. The quantities of uranium recovered by elution were generally higher, sometimes up to 10-20%, than those calculated from the loading curves. This might have resulted from inaccuracy in determination of uranium concentrations in the effluent, and, particularly, the feed. Since the uranium complexes in the feed are made unstable by lowering the pH, some uranium could precipitate resulting in lower uranium concentration in the analyses.

To determine the nature of the uranium loaded on the resin, the resin was eluted with 1 N NaCl first and then with 1 N HCl. The total uranium loaded is the sum of the uranium in the two eluates. The fraction of the uranium which cannot be eluted with 1 N NaCl but has to be "leached" out with 1 N HCl is the non-exchangeable uranium.

To improve the accuracy and assess the reproducibility of the loading experiments, two loading and elution cycles were performed on each resin at each condition. The reproducibility was within ± 5% and the average loading capacity is reported in all cases.

RESULTS AND DISCUSSION

The uranium loading curves for resins A, B and C at pH 6.5 are shown in Figure 1; the effects of leachate composition on uranium loading are shown in Figures 2 and 3; and the elution curves for resin A is shown in Figure 4. The loading capacities at 25% and 100% uranium leakage are summarized in Table 3. The loading capacity at 100% uranium leakage can be considered as the ultimate loading capacity.

Uranium Loading Capacity

As shown in Table 3, the ultimate uranium loading capacity of resin A was 12.7 lb/ft^3 using solutions containing not only 200 ppm of U_3O_8, but also about 14,000 ppm of competing ions. If the selectivity of the resin were 100% for the uranium complex, the theoretical uranium loading capacity based on ion exchange capacity would be 5.6 and 11.2 lb/ft^3 when the uranyl complexes were UTC and UDC, respectively. It is recognized that all of the uranyl complexes cannot be UDC, so that the theoretical loading capacity is lower than 11.2 lb/ft^3 even in the absence of competition. The actual loading capacity is in excess of this maximum possible loading capacity by 13%. The loading capacity for resin B was somewhat lower, but its loading capacity of 11.2 lb/ft^3 also exceeded the theoretical capacity by 4%. Such high loading of uranium was unexpected. With the high level of competing anions, particularly Cl$^-$, the expected loading capacity is about 3 to 4 lb/ft^3. Apparently, the mechanism of uranium loading on the resin was not entirely by way of ion exchange.

Nature of Uranium on the Resin

The loaded resin from Run 15 was first eluted with 1 N NaCl. A typical elution curve was obtained (Figure 5), but the maximum concentration of the eluate, 1.5 g/ℓ, was very low. Elution with 10 bed volume of 1 N NaCl produced only 0.7 lb/ft^3 of uranium, which turned out to be only 5.5% of total U_3O_8 on the resin (Figure 7). That is to say, fully 94.5% of the uranium left on the resin was not in an ion exchanged state. It is proposed that the uranium is "precipitated" or deposited on the resin as uranyl hydroxides, $(UO_2)_x(OH)_y^{(2x-y)}$ or "insoluble" uranyl carbonate UO_2CO_3.

The Proposed Loading Mechanism

By adjusting the pH of the solution to near 6.5 or lower, the uranium complexes are made unstable. This is particularly true for the solutions with low total carbonates concentration, such as leachate from an in situ leaching process. When the pH of the leachate, such as solution 1, was lowered by titration using HCl, the uranium started to precipitate at pH below 5.5 and finally redissolved again when the pH was decreased further to below 4. In this pH range, the UO_2^+ could hydrolyze to form low solubility hydroxides (9).

$$X\ UO_2^{2+} + Y\ H_2O \rightleftarrows (UO_2)_x(OH)_y^{2x-y} + Y\ H^+ \quad (3)$$

Some uranium complexes in the solution are exchanged with the Cl$^-$ anion on the resin and others precipitate on the resin in a side reaction. Obviously, this uranium precipitation could be induced and enhanced by decreasing pH locally. It is not clear what causes this pH lowering and initial precipitation or hydration to occur. It might be connected with the differences in dissociation constants of the pair of ions where Cl$^-$ and $UO_2(CO_3)_2^{-2}$ are involved. Generally, the pH of the leachate decreases somewhat after ion exchange for uranyl carbonate complex removal.

The initial precipitation will create local acidity according to Equation (3) and "auto" catalyze further precipitation. It is conceivable that this acidity can convert ion exchanged uranyl carbonate complexes into hydroxide precipitates.

An alternative explanation is that at the pH range used the uranyl complex is predominantly converted to low solubility $UO_2(CO_3)$ according to the phase diagram of $U-O-H_2O-CO_2$ system by Rich, et al. (10). Alternatively, the increased uranium loading could result from site sharing (11) according to the following reaction scheme:

$$\begin{array}{c} R^+ \\ R^+ \end{array} \Big\rangle UO_2(CO_3)_2^{2-} + H_2CO_3$$

$$\rightleftarrows \begin{array}{c} R^+ \ H \ UO_2(CO_3)_2^- \\ R^+ \ H \ CO_3^- \end{array}$$

$$\begin{array}{c} R^+ \ H \ UO_2(CO_3)_2^- \\ R^+ \ H \ CO_3^- \end{array} + UO_2(CO_3)_2^{2-}$$

$$\rightleftarrows \begin{array}{c} R^+ \ H \ UO_2(CO_3)_2^- \\ R^+ \ H \ UO_2(CO_3)_2^- \end{array} + CO_3^{2-}$$

This reaction is probably feasible in the pH range of 6-7, but is, however, not favored on the grounds that the uranium species on the resin are not ion exchangeable. On the other hand, the $H\ UO_2(CO_3)_2^{-1}$ species postulated from the site sharing reaction could be precursors for precipitation. In addition, the effects of process variables also supports the non-ion exchange loading mechanism.

Effect of Competing Ions on Loading Capacity

The uranium loading capacity was not affected by competing Cl^- and SO_4^{2-} ions. In spite of the 4-fold variation in Cl^- from 823 to 3286 ppm and 7.5-fold variation in SO_4^{2-} from 2 to 15 g/ℓ, the uranium loading capacities remained remarkably constant (Table 3). This is contrary to conventional ion exchange processes in which the loading capacity decreases as the concentration of competing ions increases, and is particularly true for high selectivity ions such as Cl^-. In typical operations, the uranium loading capacity at Cl^- level of 3000 ppm, generally falls to about 3 lb/ft^3. In order to severely test the mechanism, a solution containing 40,000 ppm of Cl^- and 200 ppm of U_3O_8 was tested. The loading capacity was found to be 10.5 lb/ft^3 (run 49). Such insensitivity to competing ions suggests that the main mechanism of uranium loading is not conventional ion exchange.

Effect of pH on Loading Capacity

The uranium loading capacity is sensitive to pH of the solution. The relative loading capacities at pH of 8.5, 6.5 and 5.0 were found to be 1, 2.3 and 2, respectively. This is in accord with the proposed scheme. Apparently, there is an optimum pH level for maximizing the loading capacity. At the high pH of 8.5, much of the uranium is in the form of a stable uranyl tricarbonate complex so that normal ion exchange is the primary loading mechanism. When the pH is decreased to 6.5, the loading mechanism is shifted to the precipitation mode. When the pH is further decreased to 5, the uranyl complex became so unstable, that it is essentially all deposited as a precipitate. The slight decrease in loading capacity from that of pH 6.5 could be due to pore mouth plugging and the subsequent decreased effect of inducing precipitation by the resin. In support of this postulate, the fractions of non-exchangeable uranium were determined. They were 19, 71, 94.5 and 96% for loading at pH 8.5, 7.5, 6.5 and 6, respectively.

In a separate study, it was found that the optimum pH level depends on the composition of the solution particularly Cl^- and HCO_3^- concentrations and this in turn depends on the stability of the uranium carbonate complexes in the solution.

Effect of Resin

If the uranium loading is mainly by way of precipitation the type of resin should have little effect on the loading capacity, and any solid surface can be employed in lieu of the resins. However, some subtle differences in the nature of resin seems to have significant effects. Under the same conditions, the loading capacities were 12.7, 11.2 and 8.5 lb/ft^3 for resins A, B and C, respectively (Table 3). When activated carbon was used at comparable conditions, the loading capacity was 2 lb/ft^3. It is possible that each resin, or solid support, reacts with the uranium complexes and induces and catalyzes uranium precipitation. It is entirely possible that the optimum loading conditions, particularly the pH of the solution, could be different for each support.

"Elution" of Uranium

The uranium hydroxides precipitate cannot be eluted with an anionic counter ion, such as Cl$^-$ alone. To remove the precipitate from the resin, leaching is needed once more. Fortunately, such leaching is easy and fast, and can be accomplished with either acid or carbonate solutions:

Acid, particularly HCl:

$$(UO_2)_x(OH)_y^{2x-y} + Y\ HCl \rightarrow X\ UO_2^{2+} + Y\ H_2O + Y\ Cl^- \quad (4)$$

Carbonate, particularly Na$_2$CO$_3$:

$$(UO_2)_x(OH)_y^{2x-y} + 3X\ CO_3^{2-} \rightarrow X\ UO_2(CO_3)_3^{4-} + Y(OH)^- \quad (5)$$

The choice of acid or carbonate elution of uranium from the loaded resin, depends on the down stream step, i.e., uranium precipitation. It also depends, to a lesser extent, on the composition of the leaching and elution circuit. For instance, if calcium ions are present, the acid solution is preferred.

The elution curve using 1 N HCl is shown in Figure 6. It eluted the precipitated uranium with high efficiency; a low volume of eluant was required, resulting in eluate of high uranium concentration, and complete elution of uranium was achieved, resulting in improved loading efficiency for the subsequent cycles. The maximum uranium concentration in the eluate was over 40 g/ℓ, and the elution was essentially complete within 5 bed volumes of eluant.

SUMMARY

A process scheme for effectively recovering uranium from dilute solution, particularly alkaline carbonate leachate from in situ leaching process is presented. In the process the pH of the alkaline carbonate leachate is lowered to its optimum level, say 6.5 with mineral acids or carbon dioxide. The leachate is then passed over an anionic ion exchange resin which is effective in inducing uranium precipitation and attaining a high loading of uranium in excess of its ion exchange capacity. The uranium-laden resin is eluted or leached with acid or carbonate solutions.

The scheme is particularly effective for recovering uranium from solutions containing high levels of anions such as Cl$^-$ and SO$_4^{2-}$.

In addition to conventional ion exchange, the uranium is loaded on the resin as a precipitate, perhaps, in the form of uranium hydroxides. As the pH of the solution decreases, more and more uranium is loaded as a precipitate until, at pH of 6.5, precipitation accounts for 94.5% of the uranium. The role of the resin in the precipitation of uranium is still obscure.

LITERATURE CITED

1. Merritt, R. C., "The Extractive Metallurgy of Uranium," Colorado School of Mines Research Institute, 1971, p. 137.

2. Blake, C. A., Coleman, C. F., Brown, K. B., Hill, D. G., Lowrie, R. S. and Schmitt, J. M., "Studies in Carbonate Uranium System," JACS 78, 5978-5983, 1956.

3. Hostetler, P. B. and Garrels, R. M., "Transportation and Precipitation of Uranium and Vanadium at Low Temperatures, With Special Reference to Sandstone-Type Uranium Deposits," Econ. Geology, 57, 137-167 (1962).

4. Lemoine, A., Contribution a L'etude du Comportement de UO$_2$ en Milieu Aqueux a Haute Temperature et Haute Pression: Unpub. Doctoral Thesis, Nancy, 111 pp. (1975).

5. Naumov, G. B., Some Physicochemical Characteristics of the Behavior of Uranium in Hydrothermal Solutions: Geochemistry, No. 2, 127-147 (1961).

6. Yan, T. Y., "Ion Exchange Process for the Recovery of Uranium," (to Mobil Oil Corp.), U.S. Patent 4,241,026 (Dec. 23, 1980).

7. Yan, T. Y., "Ion Exchange Resins of High Loading Capacity, High Chloride Tolerance and Rapid Elution for Uranium Recovery," (to Mobil Oil Corp.) U.S. Patent 4,321,838 (Jan. 26, 1982).

8. Johnson, D. A. and Florence, T. M., "Spectropic Determination of Uranium (VI) with 2-(5-Bromo-2-pyridylazo)-5-diethylaminophenol," Anal. Chim. Acta, 53, #1, 73-79 (1971).

9. Base, C. F., Jr. and Mesmer, R. E., "The Hydrolysis of Cations," John Wiley & Sons, New York, 1976, p. 178-181.

10. Rich, R. A., Holland, H. D. and Petersen, U., "Hydrothermal Uranium Deposits," Elsevier Scientific Pub. Co., Amsterdam, 1977, p. 44.

11. Helfferich, F., "Ion Exchange," McGraw-Hill, 1962, p. 148.

TABLE 1. PROPERTIES OF ION EXCHANGE RESINS

Resin	A	B	C
Type	I	II	II
Original Sample			
Ion Exchange Capacity,			
Total, meq/g	1.33	1.24	1.19
Salt Splitting, meq/g	1.21	1.15	1.15
Water Retention, %	58.9	55.8	55.2
Uranium Loading Capacity			
As UDC	11.2	10.8	10.6
As UTC	5.6	5.4	5.3
Density, g/cc	1.075	1.075	
Void, %	41.5	40.0	
Bead Crushing Strength g/bead	315	244	
Screen Analysis			
+16	--	--	13.8
-16, +20	39.3	24.3	51.3
-20, +30	42.2	45.2	32.7
-30, +40	15.6	24.3	2.0
-40, +50	2.7	5.2	0.2
-50	0.2	1.0	--
Eff. Size, mm	0.52	0.46	0.99
Uniformity Fac.	1.60	1.61	1.41

TABLE 2. COMPOSITION OF FEED SOLUTIONS

Solution	1	2	3	4
Component, ppm				
Cl^-	2,320	500	500	40,000
HCO_3^-	1,450	1,450	1,450	1,000
SO_4^-	10,140	10,140	1,350	--
Na^+	6,920	5,910	1,700	380
Mg^+	--	--	--	250
Ca^{++}	--	--	--	200
U_3O_8	200	200	200	200
pH	6.5	6.5	6.5	5.0

TABLE 3. SUMMARY OF ION EXCHANGE LOADING

		Solution				Loading Capacity $\#/ft^3$	
Run No.	Resin	Code	Na_2SO_4 g/ℓ	NaCl ppm	pH	@ 25%	@ leakage 100%
15	A	1	15	3286	6.5	10.7	12.7
14	B	1	15	3286	6.5	8.2	11.2
4R	C	1	15	3286	6.5	5.5	8.5
23	A	2	15	823	6.5	9.6	12.3
12	B	2	15	823	6.5	7.1	10.0
18	B	3	2	823	6.5	9.8	12.6
49	A	4	--	64,130	5.0	7.6	10.5

Notes:
  Bed configuration: ID = 0.4 cm; Depth = 25 cm
  Resin: Volume = 3 cc; Weight = 2 g
  Solution: U_3O_8 = 200 ppm, $NaHCO_3$ = 2 g/ℓ

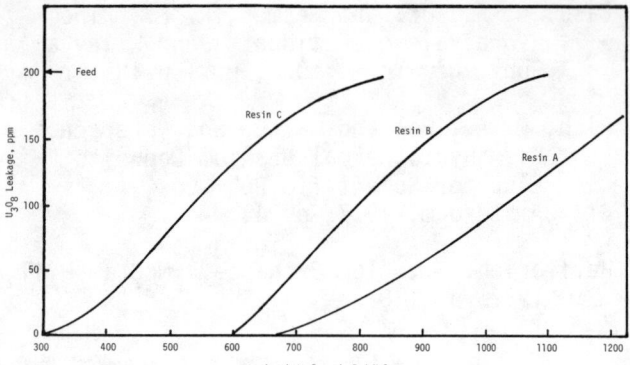

Figure 1. Uranium loading curves.

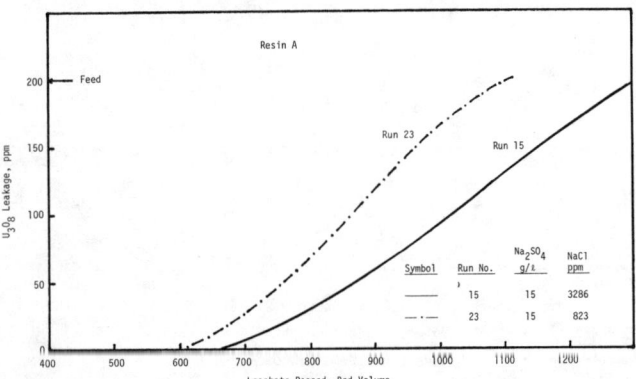

Figure 2. Effect of leachate composition of uranium leaching.

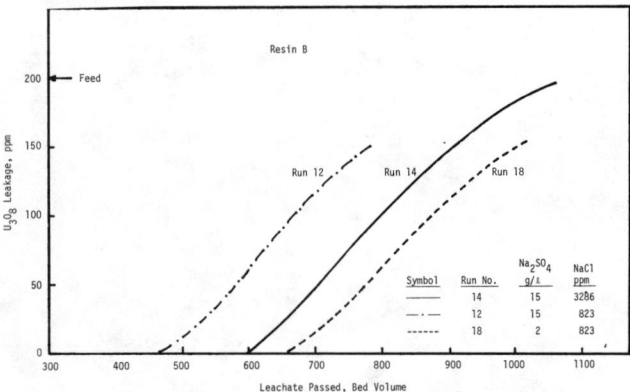

Figure 3. Effect of leachate composition of uranium loading.

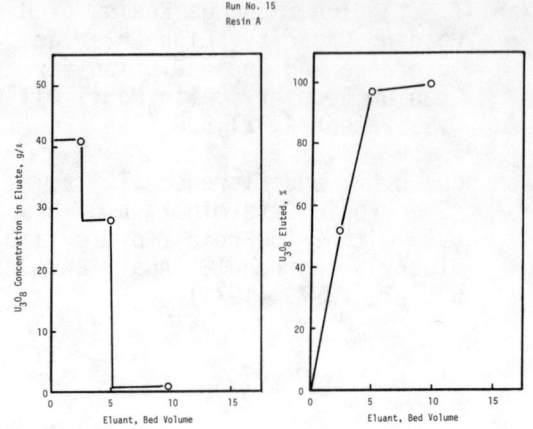

Figure 4. Elution curves.

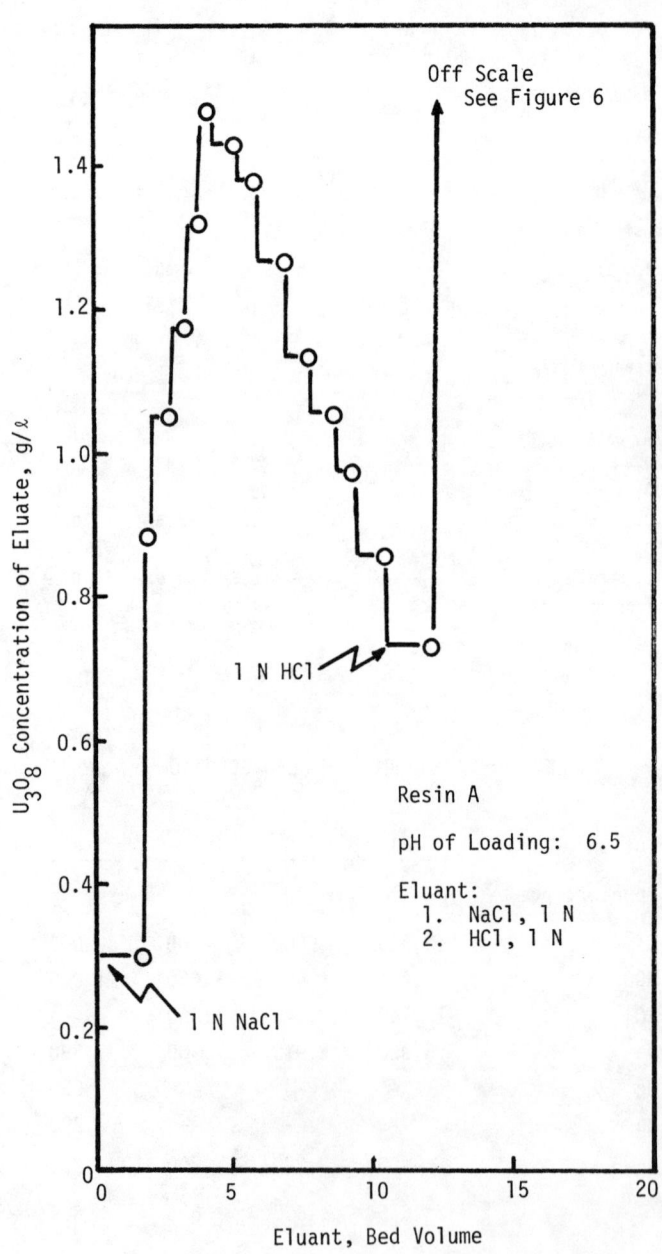

Figure 5. U_3O_8 concentration of eluate vs. bed volume of eluant.

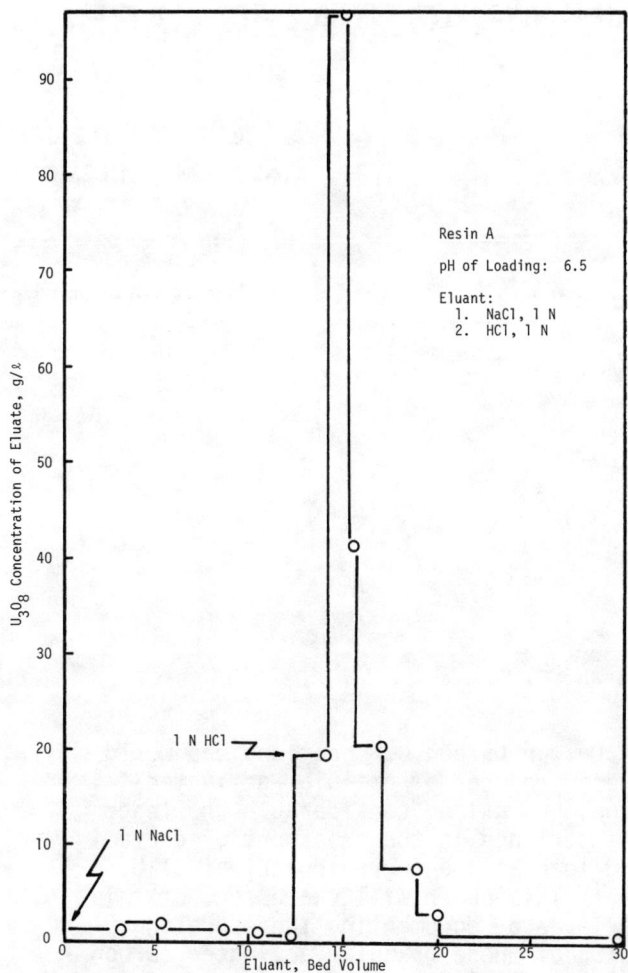

Figure 6. U_3O_8 concentration of eluate from NaCl followed by HCl.

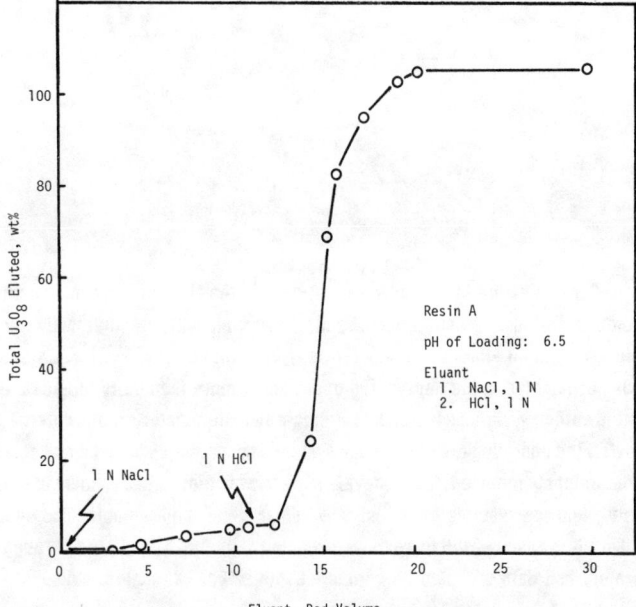

Figure 7. Cumulative elution curve with NaCl followed with HCl.

TREATMENT OF CONTAMINATED GROUNDWATER WITH GRANULAR ACTIVATED CARBON

ROBERT P. O'BRIEN
MALCOLM M. CLEMENS
and
WAYNE G. SCHULIGER
Calgon Carbon Corporation

Contamination of groundwater supplies in the U.S. with organic chemicals has become a serious widespread problem. Adsorption with granular activated carbon has proven to be an effective means for purifying contaminated waters. Calgon Corporation has participated in the design and operation of more than thirty full-scale granular activated carbon systems used in the treatment of contaminated groundwater. This paper will review the operating results of these systems, documenting the type and concentrations of contaminants removed, the removal efficiencies achieved, the amount of granular carbon used, and the systems backwash and contact time requirements. The operation of two Adsorption Systems will be reviewed in detail depicting typical breakthrough curves. Laboratory test data will also be compared with actual full-scale results.

In the United States today, 95% of all the available water supply is found under the ground. It is therefore surprising that our nation currently uses three times more surface water than groundwater. As the overall demand for water increases over the next few decades, however, we will be forced to turn to groundwater resources as our surface water supplies have largely been allocated. Today, half of the United States' drinking water comes from wells, and it is estimated that by the end of the century groundwater will supply at least 1/3 of our total national needs.

Faced with these realities, it is understandable that protection and purification of our groundwater resources is becoming a high national priority. Protection and purification will not be an easy task, however, as the EPA has identified nearly 200,000 landfills and dumpsites and 176,000 liquid disposal pits, ponds and lagoons in the U.S. which produce tens of billions of gallons of leachate each year.

Various methods of containing and treating the leachate and groundwater from the above sites will need to be employed. Granular activated carbon is one technology which over the past five years has proven to be an effective and economical method for removing organic contaminants from leachate and groundwater.

Calgon Carbon Corporation, the world's largest producer of granular activated carbon, has gained considerable knowledge and experience in the treatment of contaminated leachates and groundwater at 31 U.S. sites. This paper will share some of this experience by documenting the operating results of these granular activated carbon systems -- including performance, design requirements, and operating costs. Two sites will be reviewed in detail with attention given to system performance and operating costs.

In addition, the paper will focus on a new Calgon laboratory technique that greatly accelerates evaluation of activated carbon use for many liquid phase applications, including leachate, spill, and groundwater treatment.

PERFORMANCE-EFFECTIVE REMOVAL OF ORGANIC CHEMICALS

In our granular carbon treatment of leachates and groundwaters generated from waste storage lagoons, industrial accidents, and railroad or truck accidents, we detected a total of 27 different organic compounds. Trichloroethylene, tetrachloroethylene, dichloroethylene, trichloroethane, carbon tetrachloride, and chloroform were the compounds most frequently found. Other compounds including xylene, toluene, and benzene were also encountered. Influent concentrations of the various organic compounds ranged from the mid ug/l levels to as high as several mg/l.

As shown in Figure 1, concentrations of the compounds were consistently in the low ug/l range following carbon treatment. For the chlorinated organic compounds, the most prevalent contaminants, effluent concentrations of less than 1 ug/l (minimum detection limit) were achieved. From these results, it can be concluded that treatment with granular activated carbon is an effective technique for the removal of a wide range of organic compounds. Both virgin Calgon Filtrasorb 300 and reactivated granular carbon were used at the treatment sites.

DESIGN CRITERIA

In applying adsorption with granular activated carbon, consideration needs to be given to four key design parameters:

1. Adsorber Design Configuration
2. Pretreatment Needs
3. Contact Time Requirements
4. Carbon Usage Rate

Adsorber Design -- Downflow Fixed Bed Preferred

At all but a few of the 31 treatment sites, downflow fixed bed adsorption systems were used. The selection of the optimal adsorber design is an important consideration. Calgon has designed and operated over 250 fixed bed units, series or parallel mode, and over 150 moving bed or pulse bed systems. Our experience with both types of adsorption system designs verifies that a downflow fixed bed is more cost-effective than an upflow pulse bed for groundwater treatment applications. The adsorption wavefront (mass transfer zone) in groundwater applications is typically less than 8 feet long and can be efficiently contained in a fixed bed configuration. The mass transfer zone for groundwater is much shorter than the 30 to 40 ft. long wavefronts found in typical chemical process applications employing pulse beds. A properly designed fixed bed carbon system will operate at the same carbon usage rates as a pulsed bed system and costs less to build.

Pretreatment -- None Required At Most Sites

At one time, it was speculated that most groundwater applications would require pretreatment prior to carbon adsorption in order for the adsorption step to be properly applied. Experience has demonstrated, however, that treatment of the groundwater prior to carbon adsorption was required at only 7 of the 31 sites. In most of these cases, multi-media filtration was employed mainly as a safety factor because the quality of the contaminated groundwater, in particular suspended solids, was not fully defined prior to startup. At the majority of sites, groundwater was applied directly to the beds of granular carbon.

At three of the sites, air stripping was employed prior to carbon adsorption to reduce the level of volatile organic compounds, and to allow the carbon adsorption system to act as a final polishing unit.

Contact Time -- Minimum of 12 to 15 Minutes Recommended

The most critical design parameter for any adsorption system is contact time, which is the length of time that the contaminated water is in intimate contact with the activated carbon. In carbon systems, this important parameter is expressed as superficial contact time, which is the volume occupied by the activated carbon bed divided by the water flow rate. Since granular activated carbon systems typically have a 40 to 50% void space, the real contact time is approximately one-half of the superficial contact time.

Historically, contact times have ranged from about 7½ minutes for taste and odor removal in surface waters to several hours in some industrial processing applications.

While the contact times used in these 31 projects vary considerably, over half of the projects treating influent contaminant levels in the mg/l range had contact times of less than 1 hour (see Figure 2), and almost 60% of the systems treating influent concentrations in the ug/l range had contact times of less than 30 minutes (see Figure 3). Contact times as low as

12 minutes have been used successfully. A minimum of 12 to 15 minutes contact time is recommended for the treatment of groundwater.

The contact times used in a number of cases did not necessarily represent the "optimal contact time" for removing the particular organic compounds. In responding to emergencies such as spills, our standard or readily available adsorption equipment was used. Minimizing response time to get the system operational took precedence over the optimization of contact time.

Additionally, handling of incoming and outgoing spent carbon is a consideration. General road weight restrictions limit the weight of wet carbon that can be hauled to 40,000 lbs. (20,000 lbs. on a dry basis). In several cases, 20,000-lb. adsorbers were selected so that the spent carbon could be removed as a single containerized truckload. In 12 of the cases, a two-stage system was selected both from the standpoint of increasing carbon use efficiency and to act as an absolute safeguard against an unexpected or premature breakthrough of organic contaminants into the effluent.

Carbon Usage Rates -- As Low As 0.1 Pounds/ 1,000 Gallons

Activated carbon for organic wastewater treatment can be used on a throwaway basis, reactivated on-site, or transported for reactivation off-site. Use of carbon on a throw-away basis might be considered on potable water projects where the carbon life is measured in years. On-site reactivation is usually employed on permanent projects where the carbon requirements are large. Most of the carbon used at the treatment sites discussed in this paper was returned to Calgon's reactivation centers. This approach provides both economical carbon reuse and the total destruction of the organic compounds adsorbed on the carbon.

Fifty-three percent of the groundwater projects treating mg/l levels of contaminants used less than 1.54 lbs. of virgin carbon per each 1,000 gallons of water treated. At these projects the contaminant concentration varied from 2.0 mg/l to 200 mg/l (see Figure 4).

As expected, the carbon usage rate was lower in these projects having organic contaminants at ug/l levels. Fifty percent of the projects treating ug/l levels of contaminants used less than 0.35 lbs. of carbon per each 1,000 gallons of water treated.

OPERATING COSTS AS LOW AS 22¢/1,000 GALLONS

The costs for treating groundwater are principally dependent upon the type of equipment used and the carbon usage rate. These in turn are influenced by flow rate, concentration and type of organic compounds, type of application (potable water, etc.), site requirements, timing requirements, and length of project. Generalizations about cost, therefore, can be difficult when comparing a variety of applications and needs. Some observations, however, can be made. In general, the operating cost is lower for those projects with lower average influent levels of contaminants. The treatment cost with influent concentration at the mg/l level ranged from approximately 22¢/1,000 gallons to 55¢/1,000 gallons (see Figure 6). For those projects with influent concentrations in the mg/l range, the cost ranged from 45¢/1,000 gallons to $2.52/1,000 gallons. While these costs are highly variable depending upon the individual circumstances, the important point to be remembered is that the level of contamination was reduced to less than detectable levels in each of these cases.

CASE STUDY -- ROCKAWAY BOROUGH, NEW JERSEY

Rockaway Borough, with a population of 7,800 people, is located in northern New Jersey. Three municipal wells, each producing about 0.5 million gallons a day, supply all of the drinking water to the town's residents. Traces of tetrachloroethylene (PCE) were initially detected in one of the town's backup wells, but within 3 months levels of PCE (54 ug/l) and trichloroethylene (18 ug/l) were detected in the producing wells.

In response to the contamination, the Borough shut down the entire water system and arranged to bring water to its citizens through emergency use of tank trucks. At the same time, the Borough contacted Calgon to provide a carbon adsorption system to treat the entire 1.5 mgd water supply.

Within six weeks, three granular carbon adsorbers, each holding 20,000 pounds of virgin carbon, were installed at the

Rockaway Borough site. The adsorbers were pressure vessels operated in a downflow, parallel mode. Each adsorber treated approximately 0.5 mgd, providing a total of 15 minutes of contact time between the granular carbon and the water.

Figures 7 and 8 show the performance of the carbon adsorbers during operation in 1982. Influent levels of PCE during this period averaged 111 ug/l and were as high as 143 ug/l. Influent levels of PCE have averaged 13.8 ug/l and were as high as 19 ug/l. The treatment system has consistently produced excellent quality water.

Granular carbon in each of the vessels was on-stream for more than 10 months before meaningful levels of PCE were detected in the effluent, and the Borough made the decision to replace the carbon. Approximately 0.18 lbs. of granular carbon was used to purify 1,000 gallons of groundwater. Exhausted carbon was returned to Calgon's reactivation center where organic compounds adsorbed on the carbon were destroyed. Spent carbon was replaced on-line with virgin Filtrasorb 300 carbon, an 8x30 mesh product.

As shown in Figure 9 the annual cost for the carbon adsorption system, including equipment amortization, is $89,290 per year. At an average operating rate of 1.1 mgd, this amounts to $245/day or 21.8¢/1,000 gallons.

CASE STUDY -- TOBYHANNA ARMY DEPOT - TOBYHANNA, PENNSYLVANIA

Tobyhanna Army Depot, with a work force of 5,000 people, is located in northeastern Pennsylvania. The Army depot is a storage facility for electronic detection equipment. In mid-1981, depot water personnel discovered traces of trichloroethylene and CIS,1-2-dichloroethylene in one of their main producing wells. Within two months, levels of trichloroethylene were detected at 22 ug/l and CIS,1-2-dichloroethylene at 4 ug/l.

In response to the contamination problem, the Army closed down the well and implemented emergency water conservation methods. In August of 1981, the Army contacted Calgon to provide an emergency carbon adsorption system to treat the 0.20 MGD of water supplied by the contaminated well.

Figures 10 and 11 show the performance of this adsorption system since the installation in 1981. Influent levels of trichloroethylene and CIS,1-2-dichloroethylene have averaged 20 to 25 ug/l and 10 to 15 ug/l, respectively. The treatment system has consistently produced excellent quality water. Effluent levels of trichloroethylene and CIS,1-2-dichloroethylene have consistently been below detectable levels.

Granular carbon in the vessel was on-stream for over 14 months before the Army made the decision to replace it. Less than 0.25 pounds of granular carbon was required to purify each 1,000 gallons of groundwater treated. The used carbon was returned to a Calgon reactivation center where the organic compounds adsorbed on the carbon were destroyed. The used carbon was again replaced with virgin Filtrasorb 300 carbon.

As shown in Figure 12, the cost for the operation of the carbon adsorption system for 14 months, including equipment, maintenance, monitoring and all granular carbon needs was $37,520. Approximately 75 million gallons of water was treated during this period. The operating cost to treat 1,000 gallons of water was 50¢.

GRANULAR CARBON EVALUATION (ACCELERATED COLUMN TEST)

Time is an important factor in proving that granular carbon treatment is cost-effective -- longer carbon bed life means better economics. Carbon beds sometimes last more than a year before needing to be replaced.

Time is also an economic factor before a treatment system is ever installed. Until recently, the prediciton of carbon effectiveness in groundwater applications through pilot scale evaluations often took a few months or more. Traditional pilot scale testing techniques are expensive and time-consuming.

Responding to the need for much shorter and accurate tests to obtain design and predicted performance data, Calgon R&D developed the Accelerated Column Test (ACT). ACT is an improved technique for testing carbon's ability to remove organic impurities which combines the speed of an isotherm with the accuracy of a pilot column. The Accelerated Column Test can compress a month-long pilot study into a few days, saving money and time, which can be a critical factor in responding to spills or other emergencies.

The Accelerated Column Test generates data equivalent to a conventional lab scale or field study, which can then be easily translated into a full-scale design. Because only a few days are required to perform the tests, problems with sample degradation, or loss of volatile components, are eliminated. In fact, field validation of Accelerated Tests demonstrates a degree of accuracy that is difficult to match with any of the other more time-consuming methods.

Acceleration of the carbon adsorption cycle is achieved through a scaling down of conventional column testing hardware. Except for reduced scale, other components of the test system and the overall system design are essentially identical to larger scale laboratory or field evaluation systems (see Figure 13). The technology behind the Accelerated Column Test is based upon Calgon discoveries relating to the basic kinetics of carbon adsorption. Using kinetic data, a mathematical model of the column adsorption process was developed. With this model, breakthrough curves for full-scale adsorption systems are readily predicted from data generated by the scaled-down column.

To confirm the validity of the ACT, a series of tests were conducted comparing the ACT against a conventional one-inch column for the prediction of breakthrough for both strongly and weakly adsorbed components. For the purpose of this test, a synthetic stream was created, containing acetoxime and para-nitrophenol. As Figure 14 illustrates, the accelerated test successfully predicted the performance of the one-inch column. This degree of correlation was maintained in thirty additional tests under a variety of operating parameters.

An ACT test was also performed on a water sample from the Tobyhanna Army Depot. A 380-day operating period was simulated, treating an equivalent of 54 million gallons of water. Approximately a 2-week time period was required to conduct the test in our laboratories utilizing only 18 liters of the sample. In the test, influent levels of TCE (15 ug/l) and CIS-1,2-dichloroethylene (5 ug/l) were consistently reduced to non-detectable levels, thus duplicating the performance of the full-scale system (see Figures 11 and 12).

The ACT has been successfully employed on a number of leachate and groundwater applications and is considered an indispensible tool for evaluating the performance and economics of a granular carbon system.

SUMMARY

The problems of protecting and purifying our groundwater resources are becoming a national priority. The information presented in this paper documents the fact that granular activated carbon has an important role to play in the solution of these problems.

Treatment with granular activated carbon has been demonstrated as a cost-effective process for the removal of a wide range of organic compounds from leachates and groundwaters. Case histories have shown that non-detectable levels of synthetic organic chemicals can be achieved with total operating costs as low as 22¢ per 1,000 gallons treated and carbon usage rates as low as 0.1 lbs. per 1,000 gallons.

Also, predicting the performance of granular carbon in groundwater treatment is no longer a time consuming or expensive operation. With the development of the Calgon Accelerated Column Test, potential granular carbon users can expect a more accurate system design in less time and for less money.

	Organic Compounds In Groundwater	Number Of Occurences	Influent (1) Concentration Range	Carbon Effluent (1) Concentration Achieved
1.	carbon tetrachloride	4	130 ug/l - 10 mg/l	<1 ug/l
2.	chloroform	5	20 ug/l - 3.4 mg/l	<1 ug/l
3.	dibromochloropropane	1	2 - 5 mg/l	<1 ug/l
4.	DDD	1	1 ug/l	<.05 ug/l
5.	DDE	1	1 ug/l	<0.05 ug/l
6.	DDT	1	4 ug/l	<0.05 ug/l
7.	CIS - 1,2 - dichloroethylene	8	5 ug/l - 4 mg/l	<1 ug/l
8.	dichloropentadiene	1	450 ug/l	<10 ug/l
9.	diisopropyl ether	2	20 - 34 ug/l	<1 ug/l
10.	tertiary methyl-butylether	1	33 ug/l	<5.0 ug/l
11.	di isopropyl methyl phosphonate	1	1,250 ug/l	<50 ug/l
12.	1,3 - dichloropropene	1	10 ug/l	<1 ug/l
13.	dichloroethyl ether	1	1.1 mg/l	<1 ug/l
14.	dichloro isopropylether	1	0.8 mg/l	<1 ug/l
15.	benzene	2	0.4 - 11 mg/l	<1 ug/l
16.	acetone	1	10 - 100 ug/l	<10 ug/l
17.	ethyl acrylate	1	200 mg/l	<1 mg/l
18.	Trichlorotrifloroethane	1	6 mg/l	<10 ug/l
19.	methylene chloride	2	1 - 21 mg/l	<100 ug/l
20.	phenol	2	63 mg/l	<100 ug/l
21.	ortho chlorophenol	1	100 mg/l	<1 mg/l
22.	tetrachloroethylene	10	5 ug/l - 70 mg/l	<1 ug/l
23.	trichloroethylene	15	5 ug/l - 16 mg/l	<1 ug/l
24.	1, 1, 1 - trichloroethane	6	60 ug/l - 25 mg/l	<1 ug/l
25.	vinylidiene chloride	2	5 ug/l - 4 mg/l	<1 ug/l
26.	toluene	1	5 - 7 mg/l	<10 ug/l
27.	xylenes	3	0.2 - 10 mg/l	<10 ug/l

(1) Analyses conducted by Calgon Carbon Corporation conformed to Published U.S.E.P.A. protocol methods. Tests in the field were conducted using available analytical methods.

Figure 1. Adsorption performance.

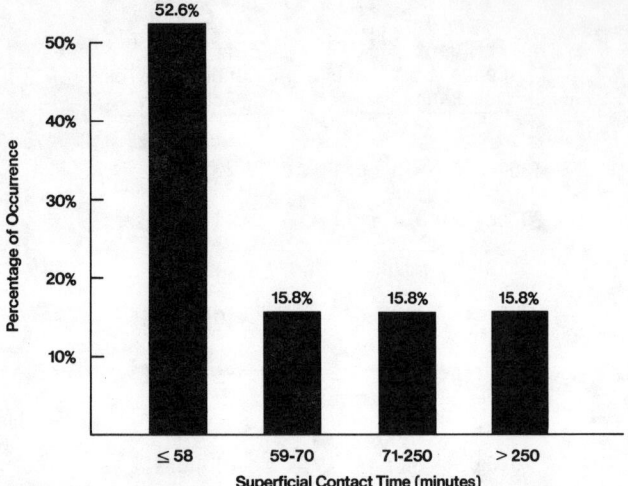

Figure 2. Range of carbon contact times for groundwater treatment projects (mg/l levels).

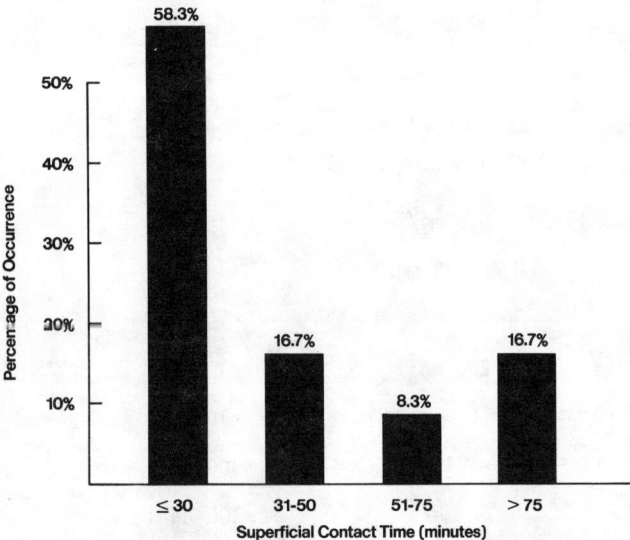

Figure 3. Range of carbon contact times for groundwater treatment projects (ug/l levels).

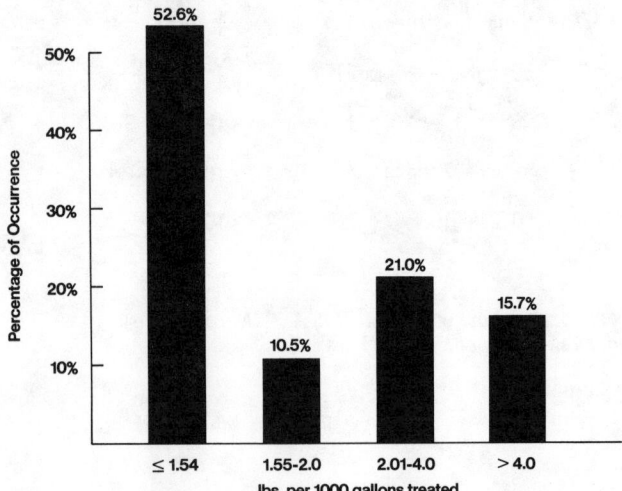

Figure 4. Range of carbon usage for groundwater treatment projects (mg/l levels).

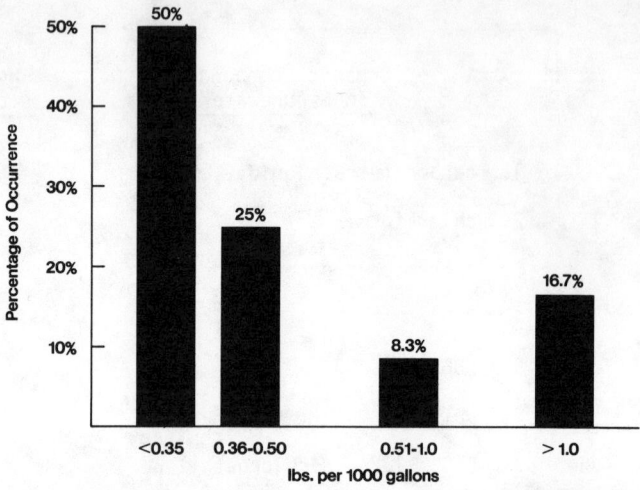

Figure 5. Range of carbon usage for groundwater treatment projects (ug/l levels).

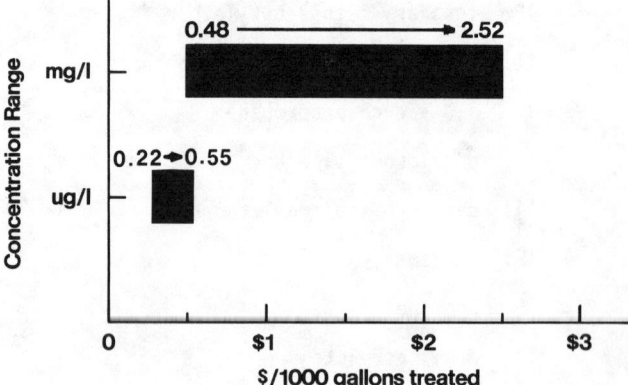

Figure 6. Granular carbon operating costs for groundwater treatment.

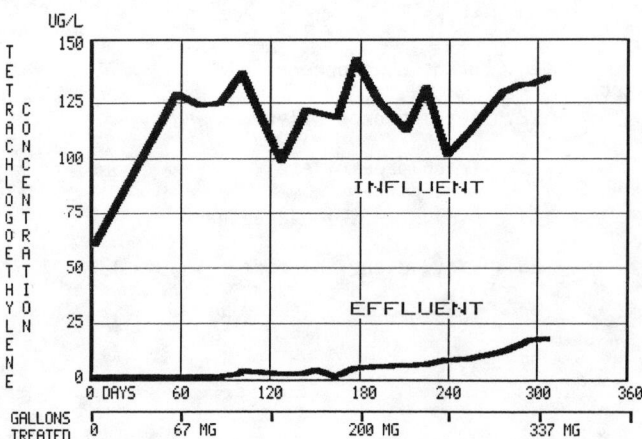

Figure 7. Rockaway Borough, N.J., granular carbon system performance.

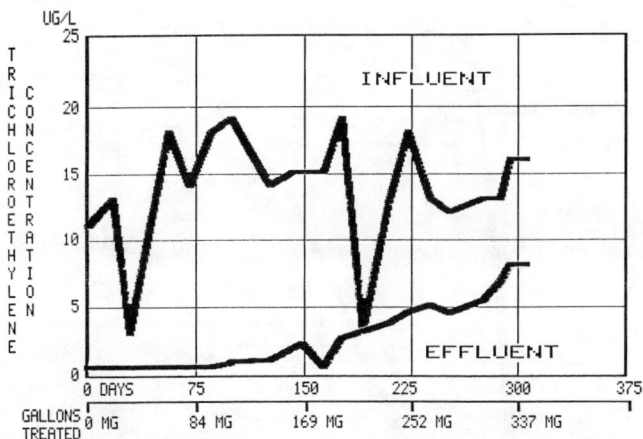

Figure 8. Rockaway Borough, N.J., granular carbon system performance.

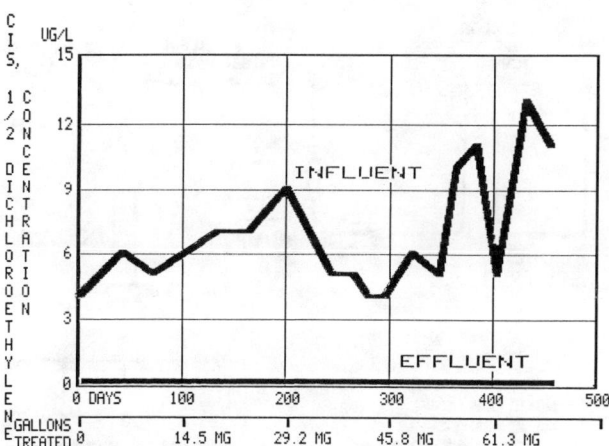

Figure 10. Tobyhanna, PA., army depot granular carbon system performance.

CAPITAL

- Adsorption Equipment, Installed Basis $136,314
  - Three 10-foot diameter pressure vessels, dished head, 14-foot straight side, 150 psig rated
  - Vessel underdrain
  - Process piping

ANNUAL OPERATING COSTS

- Virgin Granular Carbon & Monitoring $ 58,000
- Maintenance $ 6,000
- Utilities $ 1,440
- Labor $ 4,800
- Equipment Amortization (12% for 20 years) $ 18,250

Annual Total = $ 89,290

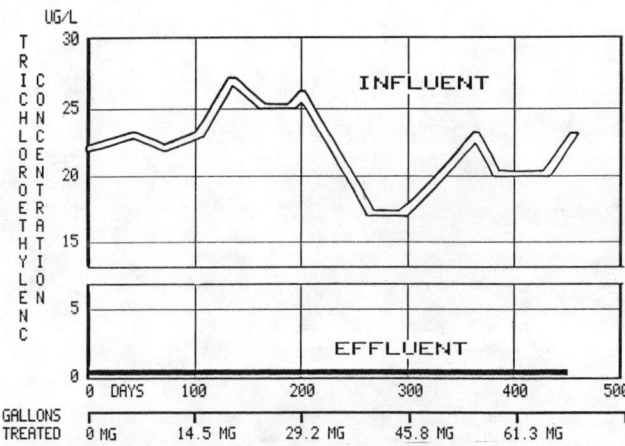

Figure 11. Tobyhanna, PA., army depot granular carbon system performance.

OPERATING COSTS (14-Month Basis)

- Service Fee (Including use of adsorption equipment, maintenance, and monitoring $17,710
- Virgin Granular Carbon $19,810

Operating Cost (14 Months) = $37,520

- Gallons Treated Per Year = 4.10×10^8
- Operating Cost Per 1,000 Gallons = 21.8¢

Figure 9. Rockaway Borough, N.J., capital and operating costs.

- Gallons Treated In 14-Month Period = 7.5×10^7
- Operating Cost Per 1,000 Gallons = 50¢

Figure 12. Tobyhanna army depot, Tobyhanna, PA.

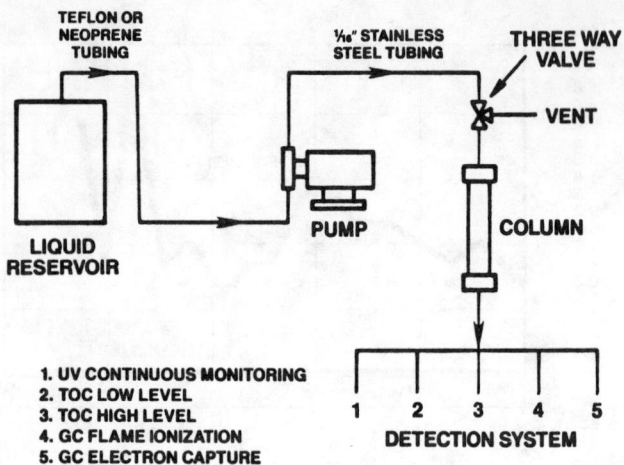

Figure 13. Basic system.

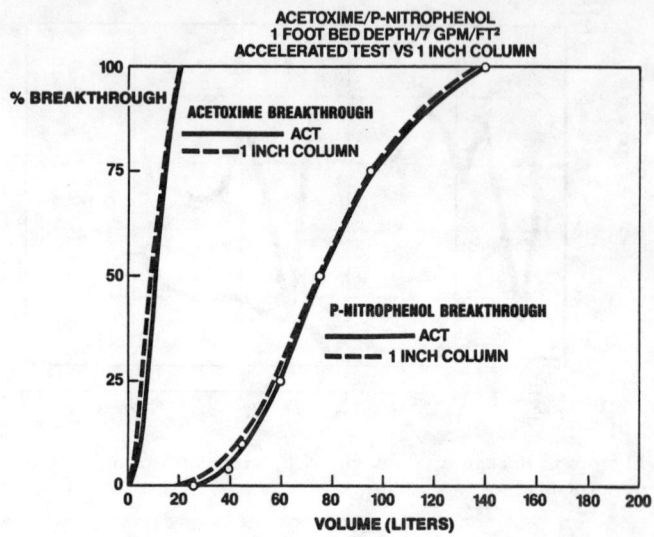

Figure 14. Multicomponent adsorption.

ADSORPTION OF ETHANOL AND WATER VAPORS BY SILICALITE

S. M. KLEIN
and
W. H. ABRAHAM

Department of Chemical Engineering and Ames Laboratory
Iowa State University,
Ames, Iowa 50011

Experimental micro-balance measurements were made of adsorption by the molecular sieve Silicalite exposed to ethanol vapor, water vapor and mixed vapors. The data are adequately represented by a Langmuir mixture model. The sieve is highly selective for ethanol from the mixed ethanol/water vapors; molar selectivities of 14 and 35 are estimated at 60°C and 25°C respectively. Predictions of ethanol adsorption from aqueous solution by Silicalite were made based on the vapor-phase data and agree well with experimental liquid-phase measurements. It is inferred that the adsorbed phase contains 98% by weight ethanol from a 10% beer at 25°C.

In the past few years, there has been renewed interest in the use of ethanol, butanol and other fermentation products as liquid fuels. A problem common to the production of these fuels is the need to separate relatively small amounts of product from large amounts of water without unreasonable energy consumption. The traditional method of distilling alcohol uses a great deal of energy; for example, recovery of anhydrous alcohol from a 10% (by weight) beer may require energy input equal to one-half of the heating value of the ethanol produced. A 2% beer may require almost 5 times as much energy input for the same amount of ethanol produced. Thus there has been a desire to find more efficient methods of separation, especially for low concentration streams. One candidate is adsorption onto a suitable solid adsorbent.

One known way to recover alcohol from aqueous solution by adsorption is to use a desiccant to remove water, but such a procedure uses excessive amounts of energy, since all of the water is later vaporized to regenerate the disiccant. For example, treatment of 10 pounds of 10% beer would require the vaporization of 9 pounds of water during the regeneration step. An energy efficient adsorption requires an adsorbent which selectively adsorbs the alcohol (minor component). One

S.M. Klein is presently with the 3M Company, St. Paul, Minnesota.

solid adsorbent which does this is the hydrophobic molecular sieve Silicalite, which is known to strongly adsorb ethanol, butanol, and many other small organic molecules from aqueous solution. Selective adsorption of organics by Silicalite has been discussed by Bibby et al. and others (3, 5, 6, 9, 12, 13 and 14).

The term 'molecular sieve' was originated to define porous solid materials which act like sieves on a molecular scale. The most important sieving effects are shown by the dehydrated crystalline zeolites, natural or synthetic, which are aluminosilicate minerals containing group I or II elements. The pore size, which can be precisely controlled, is enlarged or diminished by the incorporation of selected cations during synthesis of the molecular sieve (16).

Silicalites (SiO_2), as opposed to other zeolites are hydrophobic forms of silica because they lack sites with which water can interact. The lack of substitutional aluminum in Silicalite makes it less polar than other zeolites (2).

Flanigen et al. (6) have studied the vapor phase adsorption of several pure components by Silicalite. The amount of adsorption is reported as an equivalent liquid volume, calculated from the weight adsorbed and the density of the pure liquid at room temperature. The maximum adsorption volume for n-butane, oxygen, and methanol is reported as

0.2 cm/gm, which is interpreted as 33% void space in the Silicalite. For other molecules studied, the maximum adsorption is less. These authors suggest that adsorption in the Silicalite pores is a physical process controlled by the relative sizes of the pores and the adsorbed molecules (despite a heat of adsorption in the range of 10 to 20 kcal/mol, which might indicate chemisorption). A physical adsorption interpretation is supported by a nearly rectilinear isotherm observed for n-hexane vapor, which fills the Silicalite pores at a relative pressure (pressure/vapor pressure) of only 0.03. These authors further suggest that rejection of water by Silicalite is at least in part associated with the clustering of liquid water molecules caused by hydrogen bonding. The water clusters are thought to be too large to fit into the uniformly small pores of the Silicalite crystals.

Schumacher and Hwa (14) and others (12) measured the adsorption by Silicalite of ethanol from aqueous solutions (Figure 4). By measuring the compositions of the solutions before and after adsorption, they inferred that Silicalite adsorbs up to 0.08 kg. ethanol/kg Silicalite at low ethanol concentrations. In contrast, workers at Ames Laboratory (5) have reported adsorption as great as 0.12 kg ethanol/kg Silicalite in packed column flow experiments.

Although Silicalite clearly adsorbs ethanol strongly from aqueous solutions, attempts to recover the ethanol have been rather frustrating. Hone et al. (9) studied the separation of water/ethanol liquid mixtures by sorption using Silicalite. They reported average product composition of 12 wt% ethanol obtained by nitrogen stripping at 35°C from a column of Silicalite that had been exposed to 10% by weight ethanol in water. They concluded that a separation process using Silicalite is probably not feasible. In later studies at Ames Laboratory (5) however, workers did obtain about 75% recovery of the initial ethanol contained in a similar mixture, with a product concentration between 50 and 65% ethanol by weight.

Since Silicalite adsorbs even trace amounts of ethanol from aqueous solution, it has been presumed that the adsorbed phase contains ethanol with very little water. Nevertheless, the desorbed products have contained substantial water. One possible explanation for this discrepancy is trapping of water in larger pores and interstices between particles. In the present studies the possibility of trapped water has been minimized by measuring adsorption from a vapor phase.

EQUIPMENT AND PROCEDURES

The vapor phase adsorption measurements of these studies were made by conventional procedures with a quartz spring micro-balance and are described in detail elsewhere (10). The Silicalite used came from the Linde division of the Union Carbide Corporation. It was labeled S-115 and was in the form of small white crystals without any indication of a binder. The ethanol, water, and mixed vapors used were obtained from degassed ethanol of 99% purity and degassed water from deionized, condensed steam. Prior to adsorption measurements, the Silicalite was heated overnight at 400 to 500°C in an oven. In addition, each sample of 0.2 to 0.3 grams in the adsorption balance was evacuated by use of a standard mechanical vacuum pump for several hours at 60 to 80°C to remove impurities.

Gas phase compositions were measured with a Gow Mac series 550 gas chromatograph equipped with a Varian integrator and Porapak Q packing. Calibration of the chromatograph gave a calibration factor in good agreement with literature values (11). Reported ethanol concentrations are believed to be accurate within 1 to 2 mol %.

Each adsorption measurement was made in a similar fashion. First the Silicalite sample was evacuated for at least 15 minutes. Then it was exposed to a vapor and the weight gain noted. In general weight gain was completed within 15 to 20 seconds. Repeated measurements were made at a fixed temperature to define an isotherm. Drift of the base weight during an isotherm run was detectible but not a major problem.

Adsorption measurements were made at 60 and 115°C with vapors of pure ethanol and at 60°C with vapors containing pure water and vapors containing both ethanol and water. Results of these measurements are discussed below.

RESULTS AND DISCUSSION

Plots of the adsorption measurements are shown in Figures 1, 2, and 3. A previous study of the adsorption on Silicalite of n-hexane by Flanigen et al. (6) had shown that the isotherm was well represented by a Langmuir form, which assumes that each molecule is adsorbed independently and that the energy of adsorption is constant. We tested our data against this model.

As discussed by Adamson (1) and Breck (4), the Langmuir model can be extended to describe the adsorption of mixed vapors. Assuming that the "pore volume" parameter is the same for each component, the model may be written for ethanol/water vapor mixtures as:

$$M_e = \frac{\rho_e \cdot V \cdot b_e \cdot P_e}{1 + b_e \cdot P_e + b_e \cdot P_w} \quad (1)$$

$$M_w = \frac{\rho_w \cdot V \cdot b_w \cdot P_w}{1 + b_e \cdot P_e + b_w \cdot P_w} \quad (2)$$

where M = mass of a component adsorbed, kg/kg Silicalite

ρ = pure component liquid density, kg/m^3

V = "pore volume" of Silicalite, m^3

b = characteristic component parameter, a function of temperature, 1/kPa

P = component partial pressure, kPa

e = subscript for ethanol

w = subscript for water

The fixed coefficient V and the temperature-dependent coefficients b_e and b_w have been determined from the vapor phase adsorption data and calorimetric measurements of heat of adsorption. The values of V and b_e at 60° and 115°C were evaluated by a least-squares fit of the adsorption data for pure ethanol vapor at 60° and 115°C, shown in Figure 1. The value of b_w at 60°C was similarly computed from the pure water vapor adsorption data at 60°C, shown in Figure 2. Extrapolation of both b values to other temperatures was accomplished as described below.

The extrapolation of the coefficient b_e or b_w to other temperatures requires a value of the heat of adsorption for the corresponding pure component. Assuming a constant heat of adsorption, the temperature variation of either b_e or b_w may be described with the Clausius-Clapeyron equation as:

$$\frac{b(T_2)}{b(T_1)} = \exp\left[\frac{\Delta H_{ads}}{R}\left(\frac{1}{T_2} - \frac{1}{T_1}\right)\right] \quad (3)$$

where T = absolute temperature, K

H_{ads} = heat of adsorption for component (vapor), kcal/mol

R = universal gas constant = 1.987 kcal/mol K

For ethanol, a heat of adsorption may be inferred from equation (3) by using the fitted b_e values at 60° and 115°C. The value of 15 kcal/mol so obtained agrees with 15 kcal/mol obtained by independent calorimetric measurement. The calorimetric measurement was a simple observation of the temperature rise which results when ethanol is adsorbed from aqueous solution onto Silicalite. The inferred heat of adsorption for liquid ethanol is corrected by the known heat of vaporization to obtain a value of 15 kcal/mol for the vapor phase adsorption. For water, it is not feasible to infer a heat of adsorption from available adsorption data. As shown in Figure 2, the present measurements of pure water vapor adsorption at 60°C follow an isotherm different in form from previous measurements reported by Flanigen et al. (6) for ambient temperature. Flanigen did report a value of 6 kcal/mol for the initial isoteric heat of adsorption of water. But calorimetric measurements with liquid water adsorption in the present study showed no detectable heat of adsorption for liquid water. Accordingly, we have selected a value for the vapor phase heat of adsorption equal to the heat of vaporization for liquid water, 9.7 kcal/mol.

Parameters of the Langmuir mixture model for vapor phase adsorption are given in Table 3. The values of b at 25°C were obtained using equation (3) and heats of adsorption to extrapolate the values of b from higher temperature.

Table 3. Langmuir Mixture Model Parameters

Temperature °C	b_e kPa^{-1}	b_w kPa^{-1}	V m^3/kg Silicalite
25	**11.6	**0.102	1.52 x 10^{-4}
60	* 0.81	*0.0182	*1.52 x 10^{-4}
115	* 0.0425	--	1.52 x 10^{-4}

* By fitting pure component data
** By extrapolation

The Langmuir model has been used in several ways. In Figure 3, the predictions of the model are compared to experimental adsorption data for pure and mixed vapors at 60°C. The fit to experimental mixture data plotted was based entirely on pure component data; it is not significantly changed by a least-squares fit of all the data. It appears that the Langmuir model reasonably describes the experimental data.

Besides fitting the vapor phase adsorption data, the Langmuir model was also used to predict adsorption of ethanol from aqueous solution. If we assume the same adsorption from any given aqueous solution as from the corresponding equilibrium vapor, the ethanol adsorbed from an aqueous solution will be:

$$M_e = \frac{\rho_e \cdot V \cdot b_e \cdot x_e \cdot \gamma_e \cdot P_e^o}{1 + b_e \cdot x_e \cdot \gamma_e \cdot P_e^o + b_w \cdot x_w \cdot \gamma_w \cdot P_w^o} \quad (4)$$

where x = mol fraction of component in aqueous solution
γ = component activity coefficient
P_e^o = vapor pressure of pure ethanol.
P_w^o = vapor pressure of pure water.

The activity coefficients are calculated from formulae reported by Hansen and Miller (8), who fitted appropriate vapor-liquid equilibrium data at 25°C.

Figure 4 compares the predicted and experimental adsorption of ethanol from aqueous solution onto Silicalite. There is good agreement between predicted adsorption and values obtained at Ames Laboratory (5) using aqueous solutions flowing through packed columns at 21°C. Data of Schumacher and Hwa (14) also shown in Figure 4 indicate a lower maximum adsorption, approximately 0.08 versus 0.12 kg ethanol/kg of Silicalite. Reasons for the discrepancy are not known; however, Schumacher and Hwa do report that no pretreatment was done to activate the sieve and remove absorbed gases, which could readily account for the decreased adsorption.

Successful use of the Langmuir mixture model in prediction of ethanol adsorption from aqueous solution implies that the model correctly represents the composition of the adsorbed phase. According to the model, molar selectivity for vapor phase adsorption is a function of temperature only, since from equations (1) and (2) we find:

$$K = \frac{X_{e\ ads}}{X_{w\ ads}} \frac{Y_w}{Y_e} = \frac{MW_w}{MW_e} \frac{\rho_e}{\rho_w} \frac{b_e}{b_w} \quad (5)$$

where K = molar selectivity for ethanol
X_{ads} = component mol fraction in adsorbed phase
Y = component mol fraction in vapor phase
MW = component molecular weight
b = temperature dependent component parameter, kPa^{-1}
ρ = component liquid density, kg/m^3

The selectivity, K, calculated from equation (5) using values of the b parameters from Table 3, is 14 at 60°C and 35 at 25°C. The value of 14 at 60°C is based directly on the mixed vapor adsorption data, whereas the value of 35 at 25°C is an extrapolated value. The extrapolation, as discussed earlier, relies on numerical values of pure component heats of adsorption, and there is some question about the right value for water vapor. We used a value of 9.7 kcal/mol. If we had used the value of 6 kcal/mol suggested by Flanigen, the indicated molar selectivity at 25°C would be 65 instead of 35.

Although exact numerical values of selectivity are uncertain, these results do imply that Silicalite preferentially absorbs ethanol from mixed vapors of ethanol and water, in a manner comparable to that observed with liquid phase adsorption. This suggests that possible effects of water molecule clusters in the liquid phase (6) are not significant in determining the preferential adsorption of ethanol by Silicalite.

An alternative way to estimate the selectivity of Silicalite for adsorption of ethanol from ethanol/water vapors in the measurement of pure component retention times in a gas chromatographic column packed with Silicalite. Based on such data reported by Breck (4), we infer selectivity as equal to 6 and 10 at 200° C and 150°C respectively. Obviously these values would suggest greater selectivity at 60° and 25°C than the values of 14 and 35 reported above. We note that the gas chromatograph measurements are made at very low component concentrations, whereas the mixed vapor adsorption measurements extend to the region of nearly saturated adsorbent. Thus the differences in inferred selectivity are not entirely unexpected and further suggest that the selectivity is greater when the adsorbent is lightly loaded.

We believe the mixed vapor adsorption measurements give a reasonable measure of overall selectivity with substantial amounts of adsorption. So we have used the Langmuir model fitted to the vapor phase data to calculate adsorbed phase compositions. Figure 5 shows the predicted adsorbed phase composition as a function of temperature and ethanol content of an aqueous solution of ethanol. The inference we draw is that the adsorbed phase is on the average highly concentrated in ethanol, for example 98% ethanol by weight from a 10% beer at 25°C. At higher temperatures the adsorption is less preferential.

Unfortunately, we do not yet have direct experimental measurements of adsorbed phase composition available for comparison with these predictions.

Available desorption studies have not produced products with the expected high ethanol content, as mentioned earlier. One possible reason is the contamination of products with water trapped in larger pores and interstices of the beds of Silicalite particles. Another plausible reason is failure to get complete desorption because of strong bonding of ethanol to the Silicalite and limited rates of mass transfer. Thermo-gravimetric measurements of ethanol desorption from Silicalite have been reported by Milestone and Bibby (12) and are compared to similar measurements in in the present work in Figure 6. As seen there, the desorption begins at about 80°C but is not complete until temperature is in the range of 150° and 200°C. The thermogravimetric measurements are done with very small samples, so that mass transfer limitations are not expected. Thus these measurements suggest that the adsorbed ethanol is bonded to the Silicalite by moderately strong forces, so that desorption by simple heating may require higher temperatures and/or better mass transfer than achieved in reported desorption studies.

CONCLUSIONS

Based primarily on measurements of vapor phase adsorption of ethanol and water vapors on Silicalite, we conclude that:

1. Adsorption of ethanol/water vapors by Silicalite is adequately represented by a simple Langmuir mixture model.
2. Ethanol is preferentially adsorbed from ethanol/water vapors by Silicalite. The preferential adsorption is equivalent for vapor phase and liquid phase adsorption, when appropriate allowance is made for differences between ethanol content of an aqueous ethanol solution and the corresponding equilibrium saturated vapor.
3. Adsorbed phase fluid is inferred to be a highly concentrated but not anhydrous ethanol, for example 98% by weight ethanol from a 10% beer at 25°C.

ACKNOWLEDGMENTS

Ames Laboratory is operated for the U.S. Department of Energy by Iowa State University under Contract No. W-7405-Eng-82. This work was supported by the Assistant Secretary for Conservation and Renewable Energy through the Solar Energy Research Institute. Additional support was received from the Iowa Corn Promotion Board.

NOTATION

A	Area of chromatograph output signal, arbitrary units
b	Langmuir adsorption parameter, equation (1), kPa^{-1}
ΔH_{ads}	Heat at adsorption, vapor component, kcal/mol
k	Proportionality constant, chromatograph output, dimensionless
K	Vapor phase molar selectivity
M	Mass of component adsorbed, kg/kg Silicalite
MW	Component molecular weight
P_e	Ethanol partial pressure, kPa
P_w	Water partial pressure, kPa
P_e^o	Pure ethanol vapor pressure, kPa
P_w^o	Pure water vapor pressure, kPa
R	Universal gas constant, 1.987 kcal/mol K
T	Absolute temperature, K
V	"Pore volume" of Silicalite, m^3/kg
x	Mol fraction of component in mixture
X_{ads}	Adsorbed phase component mol fraction
Y	Vapor phase component mol fraction

Subscripts

e	Ethanol component
w	Water component
1,2	Arbitrary temperature levels

Greek Symbols

ρ	Pure component liquid density, kg/m^3
γ	Component activity coefficient, equation (4)

LITERATURE CITED

1. Adamson, A. W. _Physical chemistry of surfaces_. New York: Interscience Publishers; 1960.

2. Barrer, R. M. _Zeolites and clay minerals as sorbents and molecular sieves_. London: Academic Press; 1978.

3. Bibby, D. M.; Milestone, N. B.; Aldridge, L. P. "Silicalite-2, a silica analogue of the alumino silicate zeolite ZSM-11." Nature 280: 664-665; 1979.

4. Breck, D. W. Zeolite molecular sieves. New York: Wiley Interscience; 1974.

5. Chriswell, C. D. "Ethanol purification by adsorption." Executive Summary. Ames Laboratory Report, DOE, Iowa State University, Ames, Iowa; 1981.

6. Flanigen, E. M.; Bennett, J. M.; Grose, R. W.; Cohen, J. P.; Patton, R. M.; Kirchner, R. M.; Smith, J. V. "Silicalite, a new hydrophobic crystalline silica molecular sieve." Nature 271: 512-516; 1978.

7. Handbook of chemistry and physics. 57th ed. Cleveland: The Chemical Rubber Co.; 1976.

8. Hansen, R. S.; Miller, F. A. "A new method for determination of activities of binary solutions of volatile liquids. J. Physical Chem. 58: 193-196; 1954.

9. Hone, R. W.; Lamarchand, M.; Malaty, W. "Separation of water ethanol mixtures by sorption." Part 2. Report. Oak Ridge, Tenn.: ORNL/MIT-304; 1981. 25 pp.

10. Klein, S. M. "Adsorption of ethanol and water vapor by Silicalite, a hydrophobic molecular sieve." M. S. Thesis, Iowa State University, Ames, Iowa; 1982.

11. McNair, H. M.; Bonelli, E. J. Basic gas chromatography. 5th edition. Walnut Creek: Varian Aerograph; 1969.

12. Milestone, N. B.; Bibby, D. M. "Concentration of alcohols by adsorption on Silicalite." J. Chem. Tech. Biotechnol. 31: 1-5; 1981.

13. Schultz-Sibbel, G. M. W.; Gjerde, D. T.; Chriswell, C. D.; Fritz, J. S.; Coleman, W. E. "Analytical investigation to the properties and uses of a new hydrophobic molecular sieve." TALANTA (1982). To be published.

14. Schumacher, W. A.; Hwa, V. W. S. "Separation of water-ethanol mixtures by sorption." Part 1. Report. Oak Ridge, Tenn.: ORNL/MIT-298; 1980. 16 pp.

15. Van Ness, H. C.; Abbott, M. M. "Procedure for the rapid degassing of liquids." Ind. Eng. Fund. 17: 66-67; 1978.

16. Ward, J. W. "The nature of active sites on zeolites. Part 3. The alkali and alkaline earth ion exchanged forms." J. Catalysis 10: 34-46; 1968.

The Ames Laboratory is operated for the U.S. Department of Energy by Iowa State University under Contract No. W-7405-Eng-82. This work was supported by the Assistant Secretary for Conservation and Renewable Energy through the Solar Energy Research Institute. Additional support was received from the Iowa Corn Promotion Board.

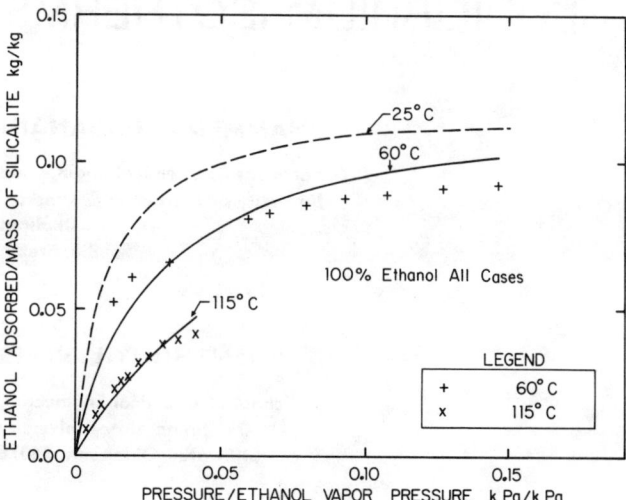

Figure 1. Adsorption of Ethanol Vapor on Silicalite—Langmuir Isotherm Fit and Extrapolation.

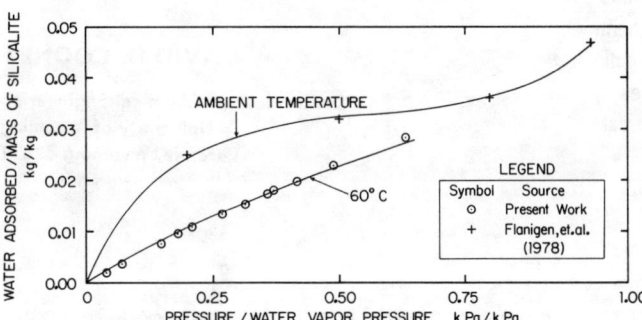

Figure 2. Adsorption of Water Vapor on Silicalite—Comparison of Present Work and Previous Work.

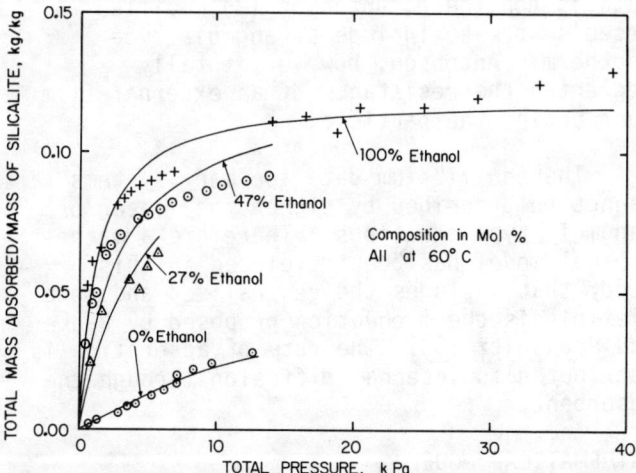

Figure 3. Adsorption of Ethanol/Water Vapors on Silicalite—Langmuir Mixture Model Fit @ 60°C.

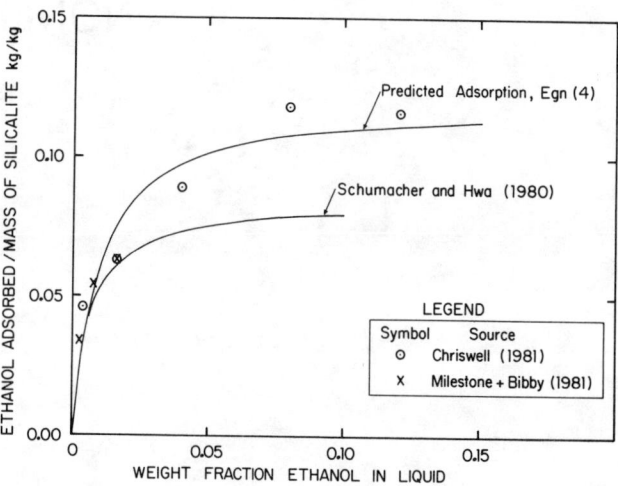

Figure 4. Comparison of Predicted and Experimental Liquid Phase Adsorption of Ethanol on Silicalite.

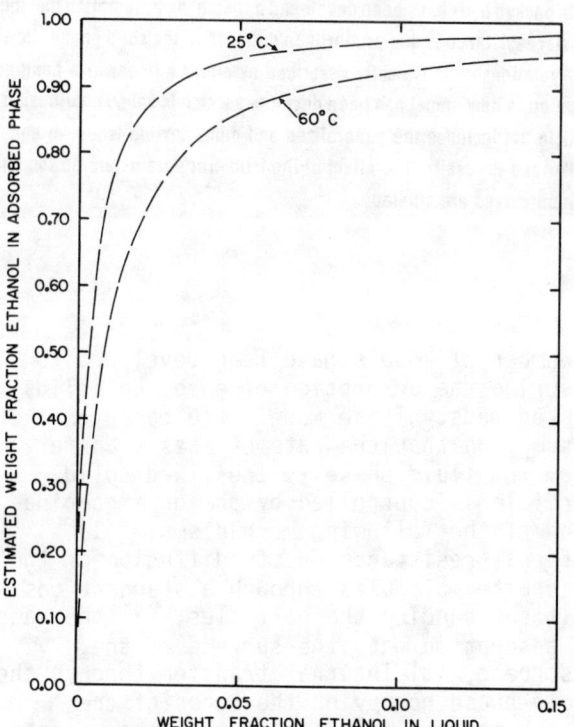

Figure 5. Estimated Composition of Adsorbed Phase on Silicalite vs. Composition and Temperature of Aqueous Ethanol Solution.

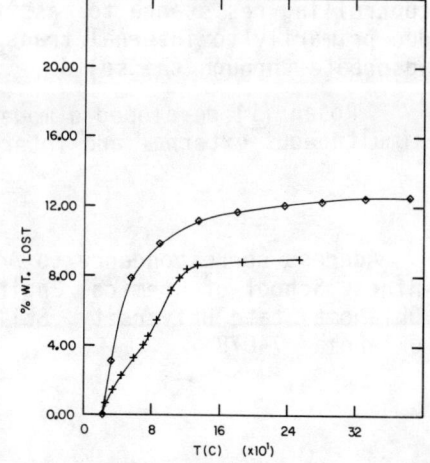

Figure 6. Desorption of Ethanol from Silicalite Determined Thermo-Gravimetrically.

ADSORPTION KINETICS FOR SYSTEMS THAT EXHIBIT NONLINEAR EQUILIBRIUM ISOTHERMS

MOHAMMED A. HASANAIN

Department of Chemical Engineering
University of Petroleum & Minerals
Dhahran
Saudia Arabia

ANTHONY L. HINES

School of Chemical Engineering
Oklahoma State University
Stillwater, Oklahoma 74078

and

DAVID O. COONEY

Department of Chemical Engineering
University of Wyoming
Laramie, Wyoming 82071

Several previous models used to describe the adsorption of gases and vapors on solids in packed beds have been developed by using the assumption that the controlling mass transfer resistance is due primarily to diffusion in the solid phase. However, the equilibrium relationship is typically described by either a linear or a Langmuir type isotherm equation. A new model has been developed which is based on diffusion through the solid particle but includes the generalized and more flexible isotherm equation proposed by Radke and Prausnitz. The effect of the isotherm parameters on the shape of the breakthrough curves are studied.

A number of models have been developed to describe the adsorption of gases on solids in packed beds. These models are based on the assumption that the rate of mass transfer from the fluid phase to the fixed solid particle is controlled by one or a combination of the following mechanisms: (1) external resistance due to diffusion of the adsorbate molecules through a stagnant gas film surrounding the particles, (2) the rate of adsorption onto the surface of the adsorbent, (3) internal transfer through the fluid phase occupying the pores of the adsorbent, and (4) diffusion of the adsorbate molecules along the surface of the pores of the adsorbent. Several investigators have concluded that for vapor phase adsorption the controlling resistance to mass transfer is due primarily to internal transport of the adsorbate through the solid.

Rosen [1] developed a model in which simultaneous external and internal mass transfer was controlling. In that model Rosen assumed that the equilibrium isotherm data could be described by a linear equation. Because a great majority of the equilibrium data is nonlinear, Antonson [2] modified Rosen's work to include a Langmuir type isotherm. Antonson, however, totally neglected the resistance of an external film surrounding the particle.

The equilibrium data for many systems cannot be described by either the linear or Langmuir type equations. Therefore a mathematical model has been developed in this study that includes the generalized and more flexible isotherm equation proposed by Radke and Prausnitz [3]. The rate of adsorption is attributed to internal diffusion through the adsorbent.

MATHEMATICAL MODEL

The adsorption of a single solute from a gas flowing through a packed bed of solids is described by

$$V_s \left(\frac{\partial C}{\partial Z}\right)_t + \rho_b \left(\frac{\partial \bar{q}}{\partial t}\right)_Z + \varepsilon \left(\frac{\partial C}{\partial t}\right)_Z = 0 \quad (1)$$

Address correspondence to Anthony L. Hines, School of Chemical Engineering, Oklahoma State University, Stillwater, Oklahoma 74078

The above equation has the following initial and boundary conditions:

$t = 0 \quad \bar{q} = 0$

$Z = 0 \quad C = C_o$

where

- C = adsorbate concentration in the fluid phase
- $\bar{q}$ = average adsorbate concentration in the solid phase
- Z = length of packed bed
- t = time
- V_s = fluid velocity in the bed based on total cross sectional area
- ρ_b = bulk density of the adsorbent
- ε = void fraction in the bed.

Based on the assumption that the adsorbent particles are approximately spherical, the diffusion through the solid is represented by

$$\left(\frac{\partial q_i}{\partial t}\right)_Z = \frac{D}{r^2} \frac{\partial}{\partial r} \left[r^2 \left(\frac{\partial q_i}{\partial r}\right) \right] \quad (2)$$

with the following initial and boundary conditions:

$t = 0 \quad q_i = 0$

$r = R \quad q_i = q_s = f(C)$

$r = 0 \quad \partial q_i / \partial r = 0$

where

- q_i = adsorbate concentration in the solid phase
- q_s = adsorbate concentration in the solid phase at the fluid-solid interface in equilibrium with C. This is obtained from the equilibrium relationship.
- R = radius of the particle
- D = effective diffusion coefficient of adsorbate through the solid.

The overall average concentration, $\bar{q}$, inside the particle is found by integrating over the particle volume as shown:

$$\bar{q} = \frac{\int_0^R q_i r^2 dr}{\int_0^R r^2 dr} = \frac{3}{R^3} \int_0^R q_i r^2 dr \quad (3)$$

In order to solve for the concentration profile in the packed bed Equations (1, 2, and 3) must be solved simultaneously. Since one of the boundary conditions for Equation (2) is a function of the unknown concentration profile, the system of equations are coupled. Equation (2) and the differential material balance given by Equation (1) can be simplified if the following change of variable is used:

$$\eta = \rho_b Z/V_s$$

and

$$\theta = t - Z\varepsilon/V_s$$

Thus Equations (1 and 2) can be expressed as

$$\frac{\partial C}{\partial \eta} = -\frac{\partial \bar{q}}{\partial \theta} \quad (4)$$

and

$$\left(\frac{\partial q_i}{\partial \theta}\right) = \frac{D}{r^2} \frac{\partial}{\partial r} \left[r^2 \left(\frac{\partial q_i}{\partial r}\right) \right] \quad (5)$$

The solution of Equation (5) for a constant surface concentration and an initial condition of zero concentration in the solid is

$$q_i(r,\theta) = 1 + \frac{2R}{\pi r} \sum_{n=1}^{\infty} \frac{(-1)^n}{n} \sin\left(\frac{n\pi r}{R}\right) *$$

$$\mathrm{Exp}\left[-D\left(\frac{n\pi}{R}\right)^2 \theta\right] \quad (6)$$

Equation (6) describes the diffusion of a solute into a spherical adsorbent particle that is initially free of solute. Since the above solution is valid for constant surface concentration it must be transformed to the case of varying surface concentration by using Duhamel's theorem. Thus

$$q_i(r,\theta) = \int_0^\theta q_s(\eta,\lambda) \frac{\partial}{\partial \theta} q_i(r,\theta-\lambda) d\lambda \quad (7)$$

The resulting equation for varying surface concentration is

$$q_i(r,\theta) = \int_0^\theta q_s(\eta,\lambda) \, 2D \sum_{n=1}^{\infty} \frac{(-1)^{n+1}}{r}\left(\frac{n\pi}{R}\right) *$$

$$\sin\left(\frac{n\pi r}{R}\right) \mathrm{Exp}\left[-D\left(\frac{n\pi}{R}\right)^2 (\theta-\lambda)\right] d\lambda \quad (8)$$

If the order of integration and summation are interchanged Equation (8) may be expressed as

$$q_i(r,\theta) = 2D \sum_{n=1}^{\infty} (-1)^{n+1} \left(\frac{n\pi}{R}\right) \frac{\sin\left(\frac{n\pi r}{R}\right)}{r} *$$

$$\int_0^\theta q_i(\eta,\lambda) \, \text{Exp}\left[-D\left(\frac{n\pi}{R}\right)^2 (\theta-\lambda)\right] d\lambda \quad (9)$$

Substituting Equation (9) into Equation (3) and integrating gives the average concentration of solute in the solid phase.

$$\bar{q} = \frac{6D}{R^3} \sum_{n=1}^{\infty} \int_0^\theta q_s(\eta,\lambda) *$$

$$\text{Exp}\left[-D\left(\frac{n\pi}{R}\right)^2 (\theta-\lambda)\right] d\lambda \quad (10)$$

Differentiation of Equation (10) with respect to θ followed by integration yields an explicit relationship for the rate of adsorbate accumulation in a particle. Thus

$$\frac{\partial \bar{q}}{\partial \theta} = \frac{6D}{R^2} \sum_{n=1}^{\infty} \int_0^\theta \frac{\partial q_s(\eta,\lambda)}{\partial \lambda} *$$

$$\text{Exp}\left[-D\left(\frac{n\pi}{R}\right)^2 (\theta-\lambda)\right] d\lambda \quad (11)$$

If the above equation is substituted into the fluid phase balance, then

$$\frac{\partial C}{\partial \eta} = -\frac{6D}{R^2} \sum_{n=1}^{\infty} \int_0^\theta \frac{\partial q_s}{\partial \lambda} *$$

$$\text{Exp}\left[-D\left(\frac{n\pi}{R}\right)^2 (\theta-\lambda)\right] d\lambda \quad (12)$$

Now q_s is eliminated by introducing the generalized equilibrium isotherm equation

$$(1/q_s) = (1/aC) + 1(/bC^\beta) \quad (13)$$

or

$$q_s = (aC/1+\alpha C^{1-\beta}) \quad (14)$$

where $\alpha = a/b$. It should be noted that Equation (14) can be readily reduced to either the linear or Langmuir equation. Upon introducing the above isotherm equation, Equation (12) can be written as

$$\frac{\partial C}{\partial \eta} = \frac{-6D}{R^2} \sum_{n=1}^{\infty} \int_0^\theta \frac{\partial (aC/1+\alpha C^{1-\beta})}{\partial \lambda} *$$

$$\text{Exp}\left[-D\left(\frac{n\pi}{R}\right)^2 (\theta-\lambda)\right] d\lambda \quad (15)$$

In order to generalize Equation (15), the following transformations are introduced:

$$\bar{Y} = C/C_0$$
$$\bar{X} = 3Da\eta/R^2 = 3Da\rho_b Z/V_s R^2$$
$$\bar{Z} = (2D/R^2)\theta = (2D/R^2)(t-\epsilon Z/V_s)$$
$$\bar{B} = \alpha C_0^{1-\beta}$$

Equation (15) thus becomes

$$\frac{\partial \bar{Y}}{\partial \bar{X}} = -2 \sum_{n=1}^{\infty} \int_0^{\bar{Z}} \frac{\partial \bar{Y}/1+\bar{B}(\bar{Y})^{1-\beta}}{\partial \lambda} *$$

$$\text{Exp}\left[-\frac{n^2\pi^2(\bar{Z}-\lambda)}{2}\right] d\lambda \quad (16)$$

with the following boundary conditions:

$$\bar{X} = 0.0 \quad \bar{Y} = 1.0$$
$$\bar{Z} = 0.0 \quad \bar{Y} = 0.0$$

Equation (16) with its associated boundary conditions can be used to describe the isothermal adsorption of a solute from a gas onto a solid desiccant. The solution of this equation for different values of α, β, and $\bar{B}$ will generate the desired theoretical breakthrough curves for this study. An explicit finite difference method was used here. A summary of the numerical scheme together with a brief discussion of the stability and convergence of the solution are given in the appendix.

RESULTS AND DISCUSSION

Solutions to Equation (16) were obtained for a variety of cases. Solutions were obtained for the case in which the isotherm equation is linear ($\beta = 1.0$) in order to check the numerical scheme and to gain some confidence in the size of the increment selected for the computations. The computed curves for the case of a linear isotherm are compared in Figure 1 to the solution of Rosen for four arbitrarily selected values of $\overline{X}$. The close agreement between the analytical solution obtained by Rosen and the computer results obtained in this study indicates the stability and convergence of the computer program for the selected increment size.

The effects of $\overline{X}$ on the shape of the breakthrough curves are illustrated in Figures 2, 3, and 4. These graphs are plotted as $\overline{Y}$ vs. $\overline{Z}/\overline{X}$ instead of $\overline{Z}$ so that the comparison of more than one breakthrough curve can be made on a single graph. The same values of $\overline{X}$ were used in Figures 2, 3, and 4 to reduce the number of variables needed to compare the breakthrough curves for the effects of the generalized equilibrium isotherm paameters. An increase in the value of $\overline{X}$ will increase the slope of the breakthrough curve and will also result in an increase in the break point time for the bed. If the variables in $\overline{X}$ are held constant except for the value of the diffusion coefficient, the previous observations are consistent with the physical situation. The increase in the value of the diffusion coefficient will decrease the resistance to internal mass transfer; hence, the bed performance will be improved.

The nonlinearity in the equilibrium isotherm is determined by the values of β, α, and $\overline{B}$. The effect of β on the shapes of the breakthrough curves is shown in Figure 5. For a constant α and $\overline{X}$ a decrease in the value of β will increase the nonlinearity in the equilibrium isotherm. The breakthrough curves become less symmetrical as the degree of the nonlinearity increases. The breakthrough curves shown in Figure 6 are plotted for a constant value of β and $\overline{X}$. The plot indicates that the curves become more nonlinear as the value of α increases. Breakthrough curves for higher value of $\overline{B}$ exhibit greater slopes in the lower part of the curve and lower slopes in the upper part than those for low values of $\overline{B}$.

ACKNOWLEDGMENT

The authors wish to acknowledge the financial support of the Solar Energy Research Institute, Golden, Colorado. The computer time necessary to make the extensive calculations for this study was provided by the Colorado School of Mines, Golden, Colorado and the University of Petroleum and Minerals, Dhahran, Saudi Arabia.

NOTATION

- a = empirical constant in the Radke-Prausnitz equation
- $\overline{B}$ = parameter characterizing breakthrough curves
- b = empirical constant in the Radke-Prausnitz equation
- C = adsorbate concentration
- C_o = inlet adsorbate concentration
- D^o = effective diffusion coefficient
- i = mesh point in $\overline{Z}$ direction
- M = $\overline{Z}/\Delta\overline{Z} - 1$, mesh point in $\overline{X}$ direction
- $\overline{q}$ = average concentration of adsorbed gas per unit mass of adsorbent
- q_i = local concentration of adsorbed gas per unit mass of adsorbent
- q_s = adsorbate concentration in the solid phase at the fluid-solid interface in equilibrium with the bulk gas concentration
- R = radius of the particle
- r = radial distance from center of particle
- t = time
- V_s = fluid velocity in bed based on total cross-sectional area
- $\overline{X}$ = $(3Da\rho_b Z)/(V_s R^2)$, effective bed length, dimensionless
- $\overline{Y}$ = C/C_o
- Z = bed height measured from column inlet
- $\overline{Z}$ = $(2D/R^2)\theta$

Greek Symbols

- α = ratio of a to b
- β = empirical constant in the Radke-Prausnitz equation
- ε = external void fraction
- ρ_b = bulk density of the adsorbent
- λ = dummy variable of integration
- η = $\rho_b Z/V_s$
- θ = $t - Z\varepsilon/V_s$, time measured from the instant the fluid reaches a point in the bed

LITERATURE CITED

1. Rosen, J. B., *J. Chem. Phys.*, 20 (3), 387 (1952).

2. Antonson, C. R., PhD, Dissertation, Northwestern University, Evantson, Ill., (1968).

3. Radke, C. J. and J. M. Prausnitz, Ind. Eng. Chem. Fund., 11, 445 (1972).

4. Antonson, C. R. and J. S. Dranoff, CEP Symp. Series, No. 74, V. 63, 61 (1967).

5. Hasanain, M. A., PhD Dissertation, Colorado School of Mines, Golden, Col., (1980).

6. Tien, C. and G. Thodos, AIChE J., 5, 373 (1959).

APPENDIX

The grid used for the solution of Equation (16) is shown in Figure A1. The development of the finite difference scheme to solve Equation (16) is carried out by following the work of Antonson and Dranoff (4). Details of the finite difference formulation are available elsewhere (5). The final difference equation for this formulation is given as

$$\overline{B}\left(\overline{Y}_{M+1}^{i+1}\right)^{2-\beta} + \left(1 + 2\Delta\overline{X}S_1\right)\overline{Y}_{M+1}^{i+1} - \overline{B}R_u\left(\overline{Y}_{M+1}^{i+1}\right)^{1-\beta} = R_u \quad (A1)$$

where

$$R_u = \left[\overline{Y}_{M+1}^i + 2\Delta\overline{X}S_1\overline{Y}_M^{i+1}\right] / \left\{1 + \overline{B}\left(\overline{Y}_M^{i+1}\right)^{1-\beta}\right\} - 2\Delta\overline{X}S_o \quad (A2)$$

and

$$S_1 = 2\sum_{n=1}^{N}\left[1 - \text{Exp}\left(-\frac{n^2\pi^2\Delta\overline{Z}}{2}\right)\right] / n^2\pi^2\Delta\overline{Z} \quad (A3)$$

$$S_o = \sum_{n=1}^{N} S_N (M-1) \quad (A4)$$

$$S_N(-1) = 0 \quad (A5)$$

The generalized equilibrium isotherm equation can be reduced to the linear isotherm equation for $\beta=1.0$. Thus, Equation (A1) reduces to

$$\overline{Y}_{M+1}^{i+1} = \frac{R_u[1 + \overline{B}]}{[1 + 2\Delta\overline{X}S_1 + \overline{B}]} \quad (A6)$$

The other limiting case of the solution can be obtained by setting $\beta=0$. The equilibrium isotherm equation then reduces to the Langmuir equation and Equation (A1) reduces to

$$\overline{Y}_{M+1}^{i+1} = \frac{-[1 - 2\Delta\overline{X}S_1 - \overline{B}R_u]}{2\overline{B}} + \frac{[(1 + 2\Delta\overline{X}S_1 - \overline{B}R_u)^2 + 4\overline{B}R_u]^{1/2}}{2\overline{B}} \quad (A7)$$

The calculations used to determine the unknown values of Y proceed as follows:

1) From known values of $\overline{Y}_1^o$ and $\overline{Y}_o^1$ calculate R_u and S_1.
2) Use Equation (A1) to calculate $\overline{Y}_1^1$.
3) Repeat step (1) using $\overline{Y}_2^o$ and the calculated $\overline{Y}_1^1$.
4) Repeat step (2) to calculate the value of $\overline{Y}_2^1$.
5) Continue in the same manner until all values of $\overline{Y}$ along the $\overline{Z}$ axis are calculated.
6) Repeat all of the above steps as the calculations move to the next set of grid points along the $\overline{X}$ axis.

The calculation begins by setting the value of the boundary and initial conditions. The boundary condition along with the $\overline{Z}$ axis is $\overline{Y}_M^o = 1.0$ for all values of M and the initial condition is $\overline{Y}_o^i = 0.0$ for all values of i along the $\overline{X}$ axis. The next step is to solve Equation (A1) to obtain the new value of $\overline{Y}_1^1$ from $\overline{Y}_1^o$ and $\overline{Y}_o^1$. At this point the scheme can move in the $\overline{X}$ direction at a fixed $\overline{Z}$ or in the $\overline{Z}$ direction at a fixed

$\bar{X}$. The latter method was selected because the result at the end of the calculations is a breakthrough curve at a fixed value of $\bar{X}$.

The solution of Equation (A1) requires the use of a root finding technique and that was accomplished by using the Regula-Falsi method. It takes a maximum of three iterations for the method to converge with the calculated root satisfying Equation (A1) up to a value of 1.0 E-7.

The question of stability and convergence is considered to be very important when dealing with explicit finite difference schemes. The development of the stability and convergence methods for the numerical solution of integro-partial differential equations is not yet well established. Tien and Thodos (6) dealing with a similar system of equations carried out an analysis that was based on physical constraints instead of sufficient conditions for stability and convergence. No attempt was made in this study to find conditions for stability and convergence. Instead, reasonable confidence in the solution was established by using the limiting cases of the generalized equilibrium isotherm and reproducing the curves generated by Rosen (1) and by Antonson (2). The increment size suggested by Antonson (2) was used for both directions. The computer program listings are given by Hasanain (5).

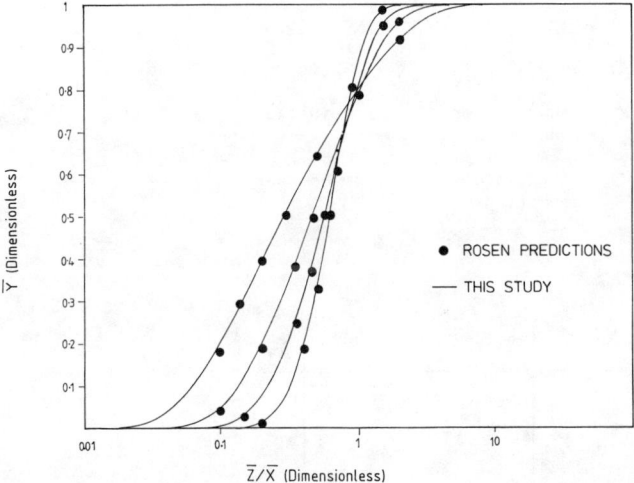

Figure 1. Comparison of predicted breakthrough curves for $\beta = 0$ and $\bar{X} = 0.2, 0.5, 1.0,$ and 2.0 with the Rosen predictions for the linear isotherm.

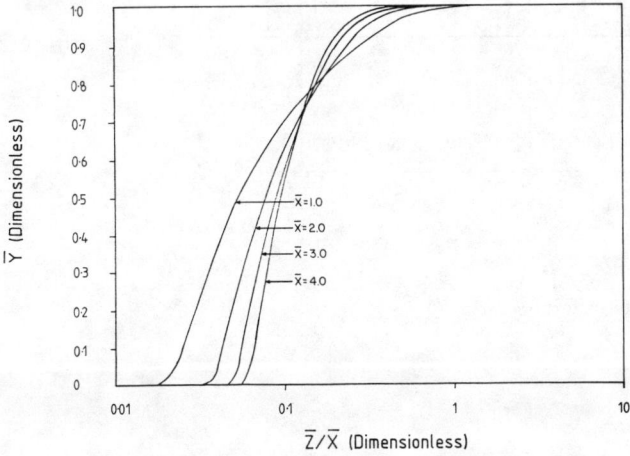

Figure 2. Graphical solution of $\bar{Y}$ vs $\bar{Z}/\bar{X}$ for $\beta = 0.21$, $\bar{B} = 4.69$.

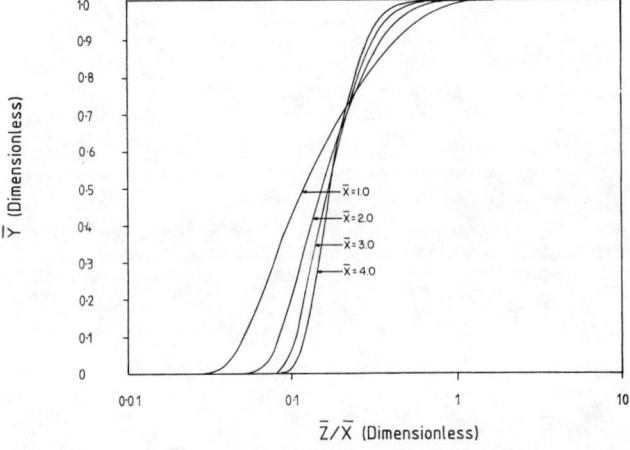

Figure 3. Graphical solution of $\bar{Y}$ vs $\bar{Z}/\bar{X}$ for $\beta = 0.305$, $\bar{B} = 2.3$.

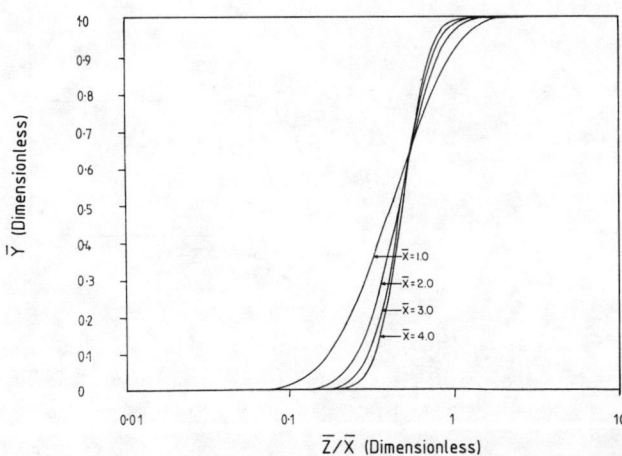

Figure 4. Graphical solution of $\bar{Y}$ vs $\bar{Z}/\bar{X}$ for $\beta = 0.24$, $\bar{B} = 0.29$.

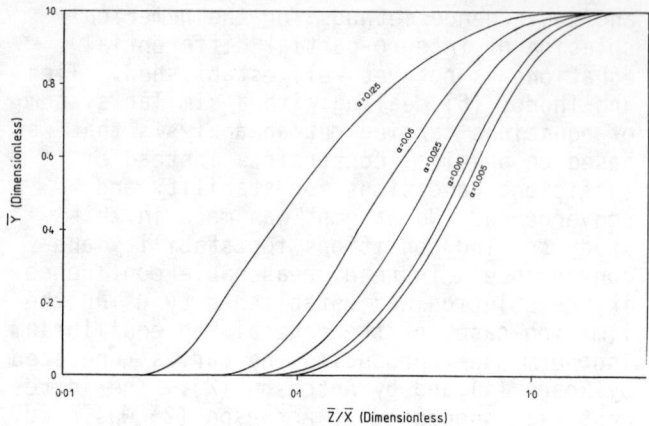

Figure 5. Effect of β on the shape of the breakthrough curves for a fixed value of α and $\bar{X}$.

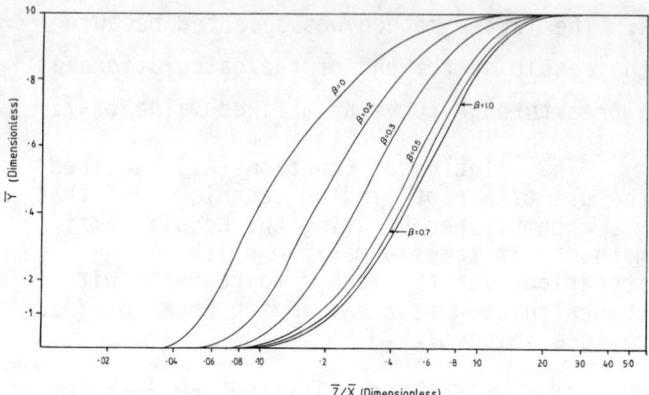

Figure 6. Effect of α on the shape of the breakthrough curves for a fixed value of β and $\bar{X}$.

○ Unknown values of $\bar{Y}$

△ Known values of $\bar{Y}$ from initial conditions

□ Known values of $\bar{Y}$ from boundary conditions

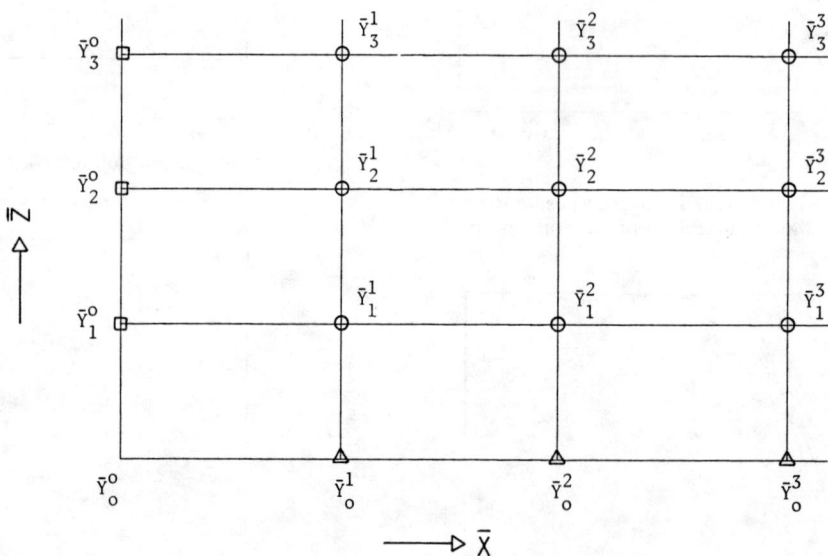

Figure A1. Finite difference grid.

HIGH PRESSURE OXYGEN, NITROGEN AND ARGON ADSORPTION IN MORDENITES

Oxygen, nitrogen and argon adsorption isotherms were measured for three mordenite samples for pressures ranging to 120 atmospheres and at 308°K to 338°K. Adsorption capacities were measured gravimetrically using a novel high pressure mass adsorption balance which allowed for continual direct measurement of adsorption bed temperature, pressure and mass as gas was introduced. These adsorption data were fitted with a high degree of success to a Dubinin-Astakhov adsorption model. The correlation was found to be general for the non-polar adsorbates oxygen and argon on the three mordenites and for nitrogen on the Na-mordenite. The adsorption potential term in the model required an additional term in order to obtain the same good correlation for the nitrogen data on Ca-mordenite. This modification was required because of nitrogen's small but finite quadrupole. Overall, the model was found to have wide applicability for the several different adsorbates on the three different mordenite samples.

DAVID T. HAYHURST
and
JENG CHENG LEE

Department of Chemical Engineering
Cleveland State University
Cleveland, Ohio 44115

INTRODUCTION

The adsorption of gases by zeolites at high pressures has received only limited attention in the published literature. This is surprising since most adsorptive and catalytic processes which use zeolites operate at pressures which are substantially above atmospheric. Recently a limited number of authors have reported high pressure adsorption data for several adsorbate/zeolite systems (1,2,3). High-pressure adsorption data have been fitted with a large degree of success to a series of generalized adsorption models. The model which has been applied most successfully is a modified form of the Dubinin-Astakhov equation. In this equation, the volumetric void filling of an adsorbent is assumed to result from the dispersive interaction of an adsorbate and adsorbent. This void filling is considered to be a direct function of the Polanyi adsorption potential for simple adsorbates. The Polanyi adsorption potential for any adsorbate is equal to the work required to liquify the molecule at the system temperature and pressure. The proportionality function is determined by the curve-fitting of experimental data. An adsorption equation which is universal for all zeolites has not yet been found, although the Dubinin-Astakhov approach has been used with considerable success for specific zeolite-types over limited temperature ranges (2).

The effects of the dispersive forces and the electrostatic fields are unique for each type of zeolitic pore system and cavity; therefore, different proportionalities between void fillings and adsorption potential should be obtained for different zeolite structures. Wakasugi et al. (2) has shown that the Dubinin-Astakhov equation can be generalized for two zeolites with similiar cation content and similar structural features. In their research, Wakasugi et al. (2) have found an identical functionality between void filling and adsorption potential for the calcium-forms of zeolites A and X.

In this study, a generalized adsorption model was determined for the adsorption of oxygen, nitrogen and argon on various forms of mordenite. The approach to developing a model for mordenite was different from that used by previous investigators (2), since in this research, the material source, the sample purity and the zeolite's cation content were varied. Two cation-exchanged forms of mordenite were tested, namely the sodium and calcium forms. In addition, the material source was also varied to include both natural and synthetic varieties of the sieve, each having different degree of purity. Data were determined for the adsorption of oxygen, nitrogen and argon into each sample at temperature ranging from 308° to 338°K and at pressure up to 1800 psia. These data were fitted to generalized form of the Dubinin-Astakhov adsorption equation in a manner similar to that used by Wakasugi et al. (2). First, a brief introduction will be given to the development of this generalized adsorption model and then the data for all gases and samples will be fitted to this model.

Generalized Adsorption Model

The generalized adsorption model used in this research is based on the work of Dubinin (4,5,6). In this original model, W_j is defined as the volume of the adsorption space occupied by the adsorbate j. It is assumed to be a function of the adsorption potential ε_j and an affinity coefficient β_j as given by the equation:

$$W_j = g(\varepsilon_j/\beta_j) \qquad (1)$$

The function g is independent of the adsorbate j, since ε_j and β_j are determined using the thermochemical properties of the adsorbing molecule. The value of the adsorption potential ε_j is given the Polanyi adsorption potential where:

$$\varepsilon_j = RT \ln (f_{s,j}/f_{e,j}) \qquad (2)$$

$$\varepsilon_j = RT \ln (f'_{s,j}/f_{e,j}) \qquad (3)$$

where $f_{s,j}$ (atm) is the fugacity of the saturated vapor of the adsorbate, $f_{e,j}$ (atm) is the fugacity of the gas in equilibrium with the adsorbed phase, $T_{c,j}$ (°K) is the critical temperature of j, R (cal/mole°K) is the gas constant and $f'_{s,j}$ (atm) is the fugacity of the adsorbate at a temperature T (T>Tc) and of a pressure P defined by $P=(T/T_{c,j})^2 P_{c,j}$ (atm) where $P_{c,j}$ (atm) is the critical pressure of j.

The evaluation of the affinity coefficient β_j requires a knowledge of the adsorptive forces acting between the adsorbate and adsorbent. Dubinin (4,5) concluded that when the adsorbent is nonpolar, the dispersive forces are predominant. This should also be true when the adsorbent surface is polar and the adsorbate molecules are nonpolar. If both the adsorbate and adsorbent are polar, then additional energys term must be introduced into the adsorption potential term to account for this enhanced affinity.

In the case where dispersive forces are predominant, the affinity coefficient is given by the equation:

$$\beta_j = \alpha_{p,j}/\alpha_{p,r} \qquad (4)$$

where $\alpha_{p,j}$ (cm^2) is the polarizability of the adsorbate and $\alpha_{p,r}$ (cm^2) is the polarizability of a reference adsorbate. In this work, argon has been chosen as the reference adsorbate where $\alpha_{p,Ar} = 1.626 \times 10^{-24}$ cm^2.

Adsorbed molecules are considered to exist in the microporous volume of a zeolite in an equilibrium state which is near to that of a liquid. This "condensation" results from the enhanced adsorption energy in the micropores due to the superposition of the fields of the opposite walls in the pores. The adsorption energy is equivalent to the energy required to "condense" an adsorbate molecule at the adsorption temperature and pressure. If this adsorption energy results from only dispersive forces, the use of the Polanyi adsorption potential for calculating ε_j is justified. If other forces such as dipole or quadrupole forces are involved, the Polanyi adsorption potential term must be modified to account for these forces.

The independent variable (ε_i/β_i) in the generalized adsorption model can be calculated for any adsorbate at any temperature. If the functional relationship between the adsorbed volume W_j and $g(\varepsilon_i/\beta_i)$ is determined for a single adsorbate over a limited temperature range, this $g(\varepsilon_i/\beta_i)$ should be general for other adsorbates and other temperatures. This characterisitic function, g, should depend only on the surface force-field of the specific adsorbent. A more detailed description of the Dubinin-Astakhov model and its application to high-pressure zeolite adsorption has been presented elsewhere (2).

EXPERIMENTAL

Material

Three zeolite samples were used in this study. They were all mordenites of either natural or synthetic origins. The first sample, called AM, was an "as mined" natural zeolite from Alaska, the second, called JM, was a processed natural zeolite from Japan and the third, called ZE, was a synthetic mordenite Zeolon 900 obtained from the Norton Company.

The Alaskan sample was obtained from Professor D.B. Hawkins of the University of Alaska. It was recieved as 10 cm. chunks of rock and was prepared for testing by simple crushing and sieving to irregular particles of less than 1.5cm in size. This larger size facilitated loading of the sample into the adsorption chamber.

The Japanese mordenite was supplied by Osaka Oxygen Industrial Co. under the trade name Zeoharb 502. It was a natural variety of mordenite. The mined zeolite had been ion-exchanged to the sodium form and fabricated into 1/16" cylindrical pellets. The Zeolon 900, the synthetic variety, was supplied by Norton Co. and was used as received in 1/8" cylindrical pellets. The molar oxide compositions of these mordenites are listed in Table 1.

Table 1. Chemical Composition of the Anhydrous Mordenites (mole %)

	AM	JM	ZE
SiO_2	81.69	81.70	82.97
Al_2O_3	9.15	9.15	8.52
MgO	1.03	0.33	0.91
K_2O	0.93	0.46	0.27
CaO	4.05	0.42	1.31
Na_2O	3.15	7.94	6.02
Si/Al	4.47	4.47	4.37

To evaluate purity, the samples were characterized using x-ray diffractometry and oxygen adsorption at $90°K$. The x-ray diffractograms were used to determine if any other crystalline phases were present. Oxygen adsorption at $90°K$ provided a measure of the true pore volume of the zeolite samples.

Adsorption runs were made with oxygen, nitrogen and argon. These gases were obtained in special high pressure cylinders in order to achieve the necessary pressure in the adsorption balance system. Each gas has had a purity of at least 99.99%. Reagent grade benzene was used in the low pressure studies to determine if the mordenite was of the large or small-pore variety. This benzene was distilled a minimum of three times to insure a purity >99.99%.

High Pressure Adsorption

The experimental apparatus used in the high pressure adsorption study is shown in Figure 1. The adsorption bed was a 4.0 cm ID cylindrical chamber fabricated from 2" nom. dia. 316SS, Schl 40 pipe and welded to 316SS end caps. The upper fitting was used to evacuate and repressurized the adsorption bed. It was fitted with a 2μm filter to prevent loss of sample during vacuuming. This upper fitting was attached to a sample line which admitted the adsorbate gas and a second line which was used for evacuation of the bed. The total system was tested under pressures up to 350 atmospheres to insure structural integrity.

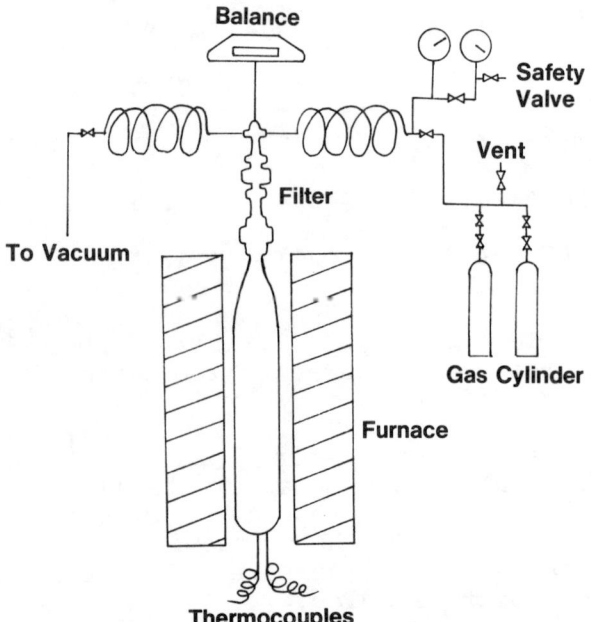

Figure 1. Schematic Diagram of the High Pressure Adsorption Balance System.

The adsorption bed was measured to have a free volume of 410 ml and large quantity of zeolite samples, up to several hundred grams, could be accommodated in this chamber. Shielded thermocouples were placed at the center and the outer surface of the adsorption chamber so that the bed temperatures could be monitored continuously. The thermocouple leads from the bed were loosely coiled so that essentially no sensitivity was lost in the weight measurement due to these lines. A furnace with two heating zones was placed around the adsorption chamber and this heater was used for both activation and run temperature control. The temperature of the bed was maintained within $\pm 1^{\circ}C$, as measured both radially and longitudinally.

The adsorbate gas line and the vacuum line were attached to the very top of the adsorption chamber using 1/8" nom. 316SS thin-wall tubing. Both of these lines were coiled as loose helicals so that external force interference was minimized. Tests revealed that the differences between weight readings taken with the lines attached and detached were statistically insignificant when compared with the sensitivity of the electronic balance ($\pm 0.01g$).

The entire sample chamber assembly was hung from a Sartorius Model 1364MP electronic balance by a hook attached to the bottom of the balance's force transducer. The balance has a capacity of 4000g and, as stated, a sensitivity of $\pm 0.01g$. The weight of the empty adsorption chamber with all lines attached was measured to be approximately 3200g. A sample of 800g could therefore be accommodated in the adsorption bed. By careful matching of the sample size with the balance's sensitivity, a high degree of accuracy was realized.

To initiate an adsorption run, several hundred grams of zeolite sample were first charged into the adsorption chamber. The vacuum line and the gas line were then attached to the adsorption chamber and the whole system was hung from the balance. To activate the zeolite sample, the gas inlet valve was closed and a vacuum of $<10^{-3}$ torr was applied. The sample was then heated to $350^{\circ}C$ and maintained at this temperature for a minimum of 12 hours. A timer switched the setting of the temperature controller from the activation temperature to a predetermined run temperature after 12 hours. The system was maintained at the preset run temperature under vacuum for five hours to insure uniform isothermal conditions in the bed. To begin the adsorption run, the vacuum line was closed, the balance was reset to zero and adsorbate gas was dosed incrementally into the adsorption chamber from a high pressure cylinder. The increments used were approximately 25 psia. The weight was monitored continuously until an equilibrium was achieved. The adsorption was considered to be at equilibrium when the system's mass change was less than ± 0.03g in 20 minutes.

Due to the large amount of the zeolite samples contained in the bed, care was exercised to insure that the bed had achieved a true isothermal equilibrium before data was taken. The heat release during an adsorption run was not found to be a significant problem. It required approximately one hour for the temperatures at the outer shell and at the center to reach the same value. This is shown in Figure 2 where temperatures were measured for a large amount of argon being dosed into a sample of AM. A maximum temperature rise of 6°K is observed. In most experimental runs, a maximum temperature rise of 2°K was observed. Temperature gradients in the radial direction were found to be negligible compared to the accuracy of the digital thermometer used ($\pm 1^\circ$C). At equilibrium, the weight and the pressure of the system were recorded.

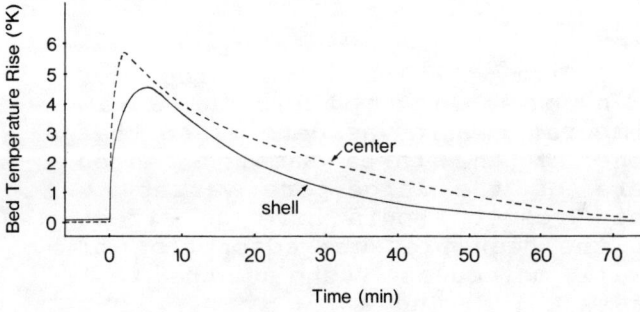

Figure 2. Adsorption Bed Temperature Rise for Ar on Alaskan Mordenite Measured at 302°K and P = 0 to 60 psia.

By subtracting the weight of the empty chamber and the weight of the sample from the recorded weight, the total weight of the gas in both free volume and in adsorbed phase was determined. The volume of the zeolite sample contained in the adsorption bed was determined by a water displacement method recommended by ASTM (7). The volume of the zeolite was subtracted from the volume of the adsorption chamber to obtain the volume of the free gas. The weight of the gas in free volume could then be calculated by an equation of state using the gas compressibility data. This weight of gas in the free volume was subtracted from the total weight in order to determine the weight of gas adsorbed. At a pressure of 1000 psi, the weight of the gas in the free volume was found to constitute 60% of the total weight of gas in the bed (adsorbed and free gas).

Low Pressure Adsorption

Adsorption isotherms were also measured at the boiling point temperature of oxygen, 90°K. This capacity was measured to determine the total pore volume of the zeolite and its corresponding purity. These measurements were made in a McBain-Bakr balance. For these runs, about 200 mg of the mordenite was placed in a hemispherical quartz pan suspended from a quartz spring. The sample was vacuumed to $<10^{-5}$ torr and further activated by heating to 300°C overnight. After activation, the adsorption chamber was cooled to room temperature and a dewar filled with liquid oxygen was placed over the adsorption chamber to jacket it completely. The level of liquid surrounding the adsorption chambers was maintained at a minimum of 15cm above the sample height to insure that the sample temperature was that of the liquid. Small amounts of gas were subsequently admitted to the adsorption chamber. The elogation of the spring was measured by a cathometer and this change in height corresponded to a change in sample weight. The adsorption isotherms were measured from 0.1 to 700 mmHg of oxygen pressure.

In order to determine if the mordenites were of the small or large pore variety, adsorption isotherm for benzene were also measured. The sample activation and adsorption measurements were identical to those described above, except the isotherms were determined at room temperature.

RESULT AND DISCUSSION

Sample Purity

The purity of the mordenite samples were determined by two methods. First, the samples were tested using x-ray diffractometry. The diffractograms indicated that each sample used in this research contained mordenite as the only zeolite phase. Both the Japanese and Alaskan mordenite samples contained large amounts of amorphous material which was found to be essentially unreacted volcanic ash. In the Alaskan sample, small amounts of quartz were also identified. No quantitative measurement was made of the percent crystallinity from the diffractograms due to the uncertainties which arose since each test sample contained different cations.

A quantative measure of the sample purity was made using the samples' oxygen adsorption at $90^{\circ}K$. At this temperatre oxygen is adsorbed as molecules in the liquid state. Since oxygen is a relatively small molecule (3.46A kinetic diameter) it can penetrate into mordenite's entire pore structure. The oxygen does not interact with the zeolite's framework and the maximum oxygen loading corresponds to a complete filling of the intracrystalline voids. Since oxygen will not be appreciable adsorbed by non-zeolite materials, the quantity of oxygen adsorbed corresponds directly to the amount of zeolite present in the sample. Other investigators have discussed the applicability of this technique for measuring a molecular sieve's purity (8).

The adsorption isotherms for oxygen at $90^{\circ}K$ on the three mordenite samples are shown in Figure 3. The different symbols for each curve corresponded to the results from different experimental runs. This indicats that the experiments were satisfactorily reproducible (within $\pm 5\%$). Data from these runs were fitted to the Langmuir isotherm equation. It was found that the Langmuir model was valid for all three samples with a correlation factor of >0.998. Listed in Table 2 is the maximum Langmuir loadings for each material, its O_2 loading at 700 torr and the relative sample purity based on the literature value for the O_2 loading of a pure Na-Zeolon at 700 mmHg and $90^{\circ}K$ as reported by Takaishi et al. (9). From these oxygen adsorption data, the void volumes of the three samples were found to be 0.137 cc/g, 0.105 cc/g and 0.0732 cc/g for Zeolon 900, Japanese and Alaskan mordenite respectively.

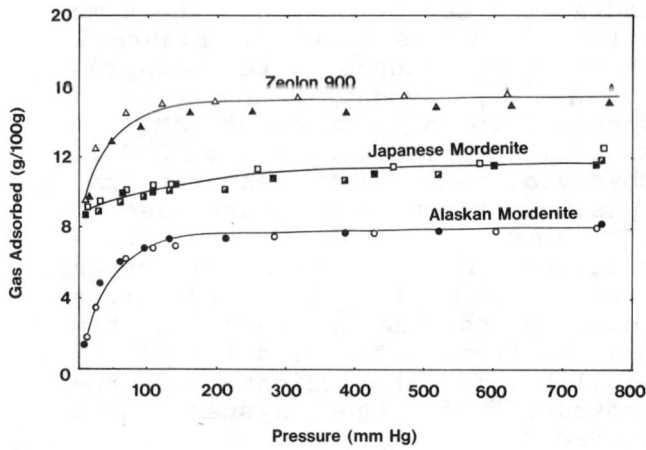

Figure 3. Oxygen Adsorption Isotherms at 90°K.

From the data measured for benzene adsorption at room temperature, it was determined that none of the three samples tested were of the large-pore variety. This result should have no effect on the high pressure adsorption of small molecules such as the test gases O_2, N_2 and Ar.

High Pressure Adsorption

High pressure adsorption isotherms for argon, oxygen and nitrogen on each of the three mordenite samples are shown in Figures 4, 5 and 6. The isotherms ranged in temperatures from 308°K to 338°K and at pressures up to 120 atmospheres. Data from each isotherm were fitted to a Langmuir model. Since experimental pressures were not high enough to see a leveling-off of the adsorption isotherms, the maximum loading of gases on these samples was determined from the Langmuir adsorption model. The Langmuir equation was used in all cases due

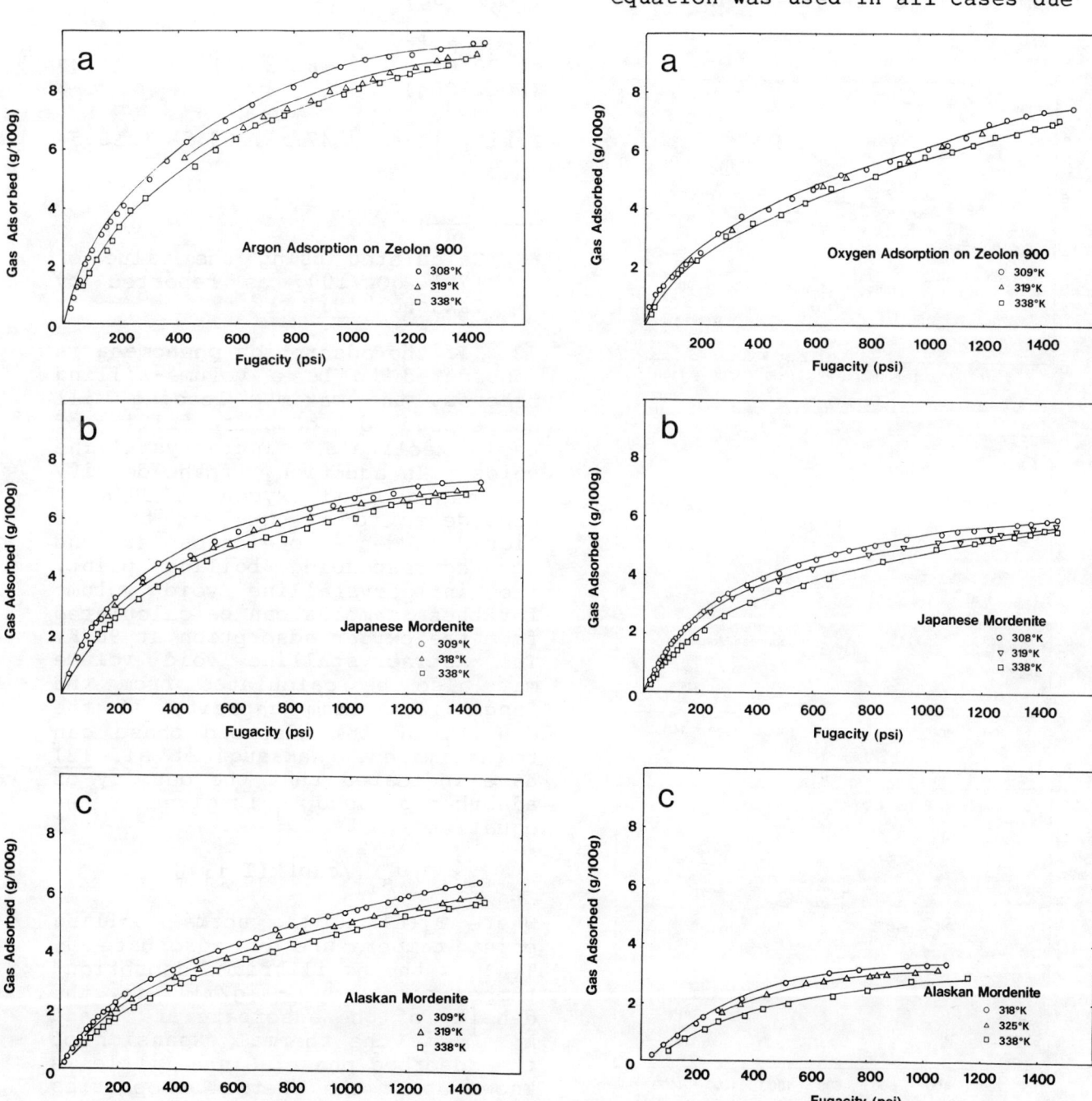

Figure 4. High Pressure Adsorption Isotherms for Argon.

Figure 5. High Pressure Adsorption Isotherms for Oxygen.

to the ease of calculation. For each sample, the calculated maximum was no more than twenty percent greater than the highest capacity measured and therefore the use of this adsorption model was considered reasonable. The maximum amounts of adsorption are considered reliable to three significant figures. These maximum loadings are listed in Table 3.

Table 2 Maximum Adsorption Capacity of the Mordenite Samples

	AM	JM	ZE
Langmuir Maximum (gO_2/100g)	8.36	11.95	15.63
O_2 Capacity at 90°K (gO_2/100g)	8.13	11.89	15.47
Purity (%)*	47.5	69.5	90.5

---

* Calculated using the value of 17.1 gO_2/100g as reported by Takasugi[9]

If the adsorption phenomena is considered to be a volume-filling process, the maximum loading will correspond to a complete filling of the zeolite's intracrystalline voids. In addition, if the density of the adsorbed oxygen at 90°K is considered to be equal to that of liquid oxygen at one atmosphere and its corresponding boiling point, the intracrystalline void volume for these samples can be calculated from the oxygen adsorption at 90°K. The intracrystalline void volume may also be calculated from the Langmuir maximum capacity if the density of the adsorbed phase can be estimated. Wakasugi et al. [2] have indicated that the density of adsorbed phase $d_{s,j}$ is given by the equation

$$d_{s,j} = d^*_{s,j}/\exp[k(T-T_j)] \qquad (5)$$

where $T_j(K)$ is the normal boiling point temperature of adsorbate j, $T(K)$ is the equilibrium adsorption temperature, $d_{s,j}$ (g/ml) is the density of the adsorbate at T_j and $k(K^{-1})$ is the thermal expansion of the adsorbed phase. Since little is known about the thermal properties of an adsorbate entrapped in the pores of a zeolite, k was approximated by the mean thermal expansion of the superheated liquid ($k=2.5 \times 10^{-3}$ °K^{-1}).

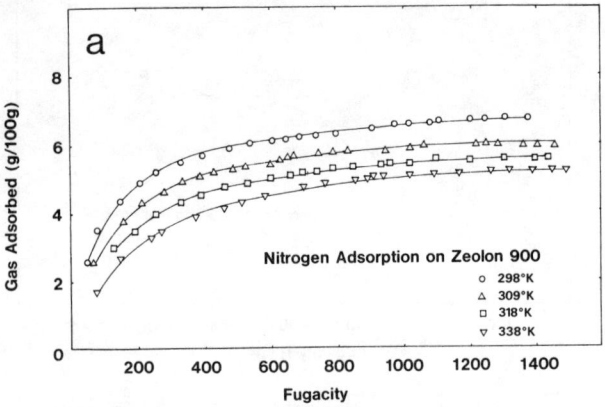
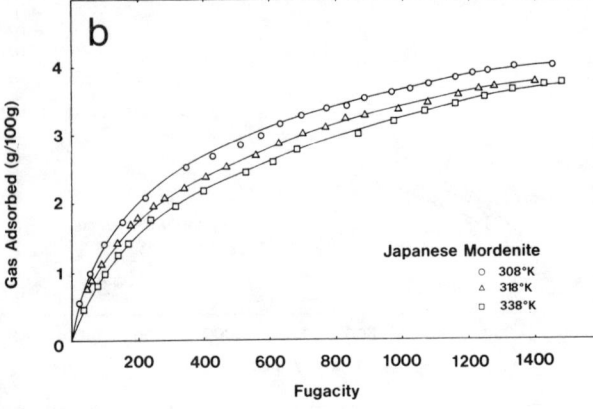
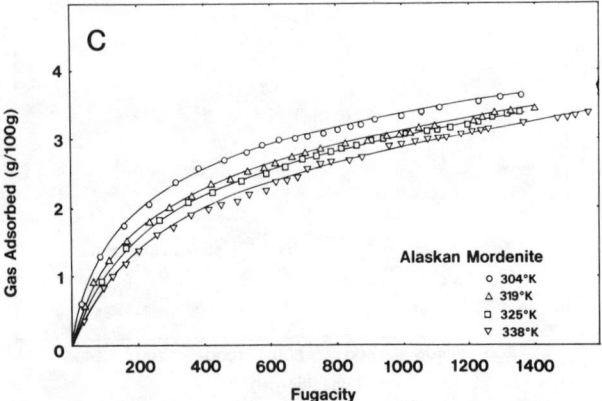

Figure 6. High Pressure Adsorption Isotherms for Nitrogen.

Table 3 Maximum Void Volume, W_o, Calculated from The Langmuir Maxima

	AM	JM	ZE
Ar*	8.2	10.3	13.3
O_2*	8.4	11.1	13.3
N_2	9.0	8.6	11.1
O_2 (at 90°K)**	7.32	10.50	13.70

* Determined from data for each gas at 308°K
** Determined from data for oxygen adsorption at 90°K

The void volumes, Wo, calculated by the two techniques outlined above are listed in Table 3. These values for Wo calculated by the two methods should be in reasonable agreement if their inherent assumptions are valid. Good agreement was found between the volume calculated from Langmuir maxima for oxygen and argon and that calculated from adsorption data for O_2 at 90°K. The maximum Langmuir volume, W_o, calculated for nitrogen was found to be different than the values of W_o calculated for oxygen and argon. This value for nitrogen was found to be substantially less than the other two values. Differences were attributed to the method used to calculate the density of the adsorbed nitrogen phase. The actual adsorbed phase density of nitrogen is expected to be greater than that calculated using Equation (5) due to the small but finite quadrupole of the nitrogen molecule.

Generalized Adsorption Model

Using the high pressure adsorption data, a generalized model based on the Dubinin-Astakhov equation was determine for mordenite samples. A plot of the logrithm of the relative void filling, W/Wo, versus (ϵ/β) was constructed for all three mordenites using the data for oxygen and argon. This plot is shown in Figure 7. The maximum void volume W_o used in these calculations was determined from the oxygen adsorption at 90°K. This choice was arbitrary and did not affect the shape of the final curve. The values of the adsorption potential, ϵ_j, were calculated from fugacity data. The values of the polarizability used to calculate the affinity coefficient, β_j, were 1.56 for oxygen and 1.626 for argon. In all calculation of β_j, argon was chosen as the reference adsorbate. Data for both adsorbates on the three adsorbent samples and at various temperatures is well represented by a single line. The use of the Dubinin-Astakhov correlation was considered to be quite reasonable for O_2 and Ar since the dispersive forces are in fact the dominant adsortive force.

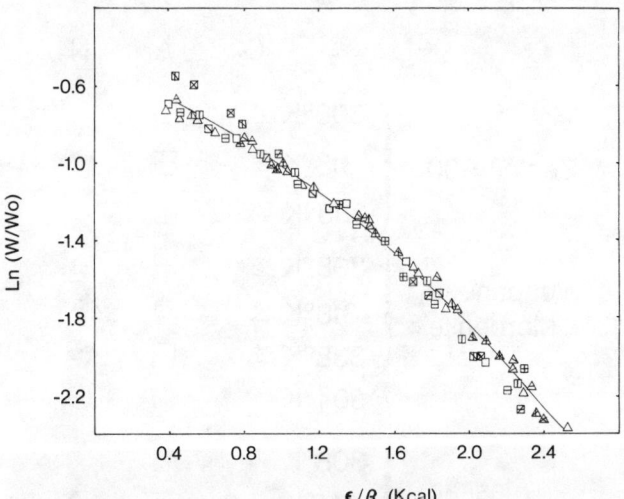

Figure 7. Characteristic Adsorption Curve for Oxygen and Argon on all Mordenites.

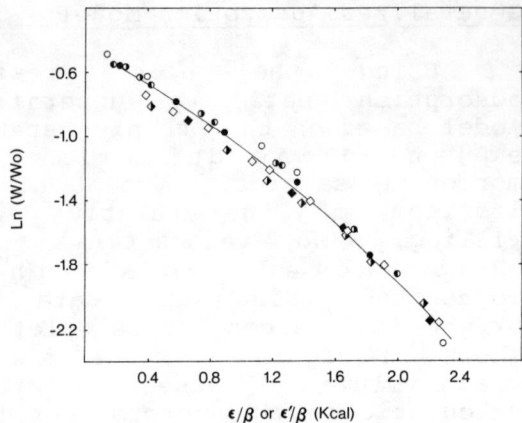

Figure 8. Characteristic Adsorption Curve for Nitrogen on the Alaskan and Japanese Mordenites.

For nitrogen adsorption, a plot of $\ln(W/W_o)$ versus (ε/β) is shown in Figure 8. For the plotting of the Alaskan mordenite data, the ε was modified by changing the adsorption potential term from ε to ε' where

$$\varepsilon' = \varepsilon - Q \qquad (6)$$

and Q is a constant. A value of Q equal to 0.26 kcal resulted in the nitrogen data for the Alaskan mordenite matching that of the Japanese sample. This corrected nitrogen curve was also found to agree quite well with the data measured for oxygen and argon.

Figure 9. Symbols Used for the Different Temperatures in Figures 7 and 8.

The need for modification of the adsorption potential term appears to result from differences in the type of cation contained in the Japanese and Alaskan mordenites. The Japanese mordenite has sodium as the dominant cation while calcium is the dominant cation in the Alaskan material. Since the energy due to dispersion, repulsion and polarization can be considered the same for the two samples (10), the quadrupole interaction between the nitrogen molecule and the divalent calcium cation must be significant. The adsorption potential which corresponds to the energy required to condense a molecule within the zeolite's pores, should account for contributions both from the dispersive forces and the quadrupole interaction. If the energy term for the quadrupole is subtracted from the total adsorption potential term, the resulting adsorption energy term should reflect those forces which are due only to the dispersive forces. This dispersive force term should be universal for nitrogen adsorption on all mordenites. With this modification of the energy term, the model was found to be general for both cation forms of mordenite. In addition, Figure 8 coincides with the Ar and O_2 data in Figure 7. This agreement between the three gases makes the correlation truly universal.

Figure 8 for nitrogen represents results from the two natural samples only. Data for the Zeolon 900 sample was not included in this plot since different curves were found for the data at each different test temperature. None of the Zeolon 900 data agreed with the data obtained for the Alaskan and Japanese mordenites except at high adsorbate loadings. The reasons for this require further investigation.

For the generalized adsorption model, no specific equation was developed which represents the curve in Figure 7. The form of this type of equation should have a fundamental significance, however, no such equation can be postulated without a specific knowledge of the adsorbate/adsorbent interaction. Current chemical and adsorption theory is still inadequate to predict such specific interactions in the complicated electrostatic fields of a zeolite. Deriving a specific adsorption equation for mordenite would be an exercise in curve-fitting if the form of the equation is not postulated from first principles. The generalized model is therefore represented only by the plot of data in Figure 7.

CONCLUSIONS

The primary objective of this research was to develop a general model for the adsorption of oxygen, nitrogen and argon on mordenite. For oxygen and argon, the dispersive forces dominate the adsorbate/adsorbent interaction. This assumption is synonymous with the derivation of the Dubinin-Astakhov equation. The model was found to yield an excellent agreement for both oxygen and argon adsorption on several mordenite samples over the range of temperatures tested. The correlation was universal for two different cation exchange forms of the mordenite. This was expected since dispersive forces are independent of the type of zeolite cation.

The generalized correlation was also found to be appropriate for nitrogen adsorbed on the sodium-form of mordenite. In the case where calcium was the dominant cation, an additional interactive forces arising from the electrostatic field-induced quadrupole had to be considered. This required a modification of the adsorption potential term to separate the contribution from dispersive forces from the quadrupole interaction. With this modification, a good correlation was found for the nitrogen data. The nitrogen curve was also found to agree well with the curve measured for oxygen and argon.

In general, surprisingly good agreement was found for the three samples tested. Each of the three mordenites was different both in total cation content and in the materials source. Each sample also contained a varying proportion of non-zeolite material. Despite these differences, deviation from the generalized adsorption curve for these materials were minimum. It can be concluded from these results that the Dubinin-Astakhov model is an excellent choice for modelling zeolite adsorption in systems where dispersive forces predominate.

ACKNOWLEGMENTS

The authors would like to acknowledge Mr. Daniel Rusnyiak of Cleveland State University and Mr. Robert Danielson of Alloy Bellow Inc. of Cleveland, Ohio for their assistance in the design and fabrication of the high pressure adsorption system. The assistance and suggestions provided by Dr. Thomas J. Walsh are also gratefully acknowledged.

The authors would also like to thank in particular the Standard Oil Company of Ohio (Sohio) for providing discretionary funding to the Chemical Engineering Department at CSU. These funds have supported this project in total.

REFERENCES

1. Takaishi, T., Yusa, A., Ogino, Y., and Ozawa, S., J. Chem. Soc. Faraday Trans. I, 70, 671 (1964)

2. Wakasugi, Y., Ozawa, S., and Ogino, Y., J. Colloid Interface Science, 79, 399 (1981).

3. Zuech, J.L., Hines, A.L., and Sloan, E.D. Ind. Eng. Chem. Process Des. Dev., 22, 172 (1983).

4. Dubinin, M.M., "Catalysis and Chemical Kinetics" Balandan, A.A. (Ed.) p. 17, Academic Press, New York (1964).

5. Dubinin, M.M. J. Colloid Interface Science, 23, 487 (1967).

6. Dubinin, M.M., Chemical Reviews, 60, 235 (1960).

7. Alder, W.R., Am. Ceram. Soc. Bull. 54, 638 (1975)

8. Bolton, A.P., "Experimental Methods in Catalytic Research" Vol. II, Anderson, R.B. (Ed.) p. 11, Academic Press, New York (1976).

9. Takaishi, T. and Yusa, A., Trans. Faraday Soc. 67, 3565 (1971).

10. Breck, D.W., "Zeolite Molecular Sieves" p. 667, John-Wiley & Sons, New York (1974).

COMPARISON OF PERFORMANCE OF PACKED AND SEMIFLUIDIZED BEDS FOR ADSORPTION OF TRACE ORGANICS

ALEXANDER P. MATHEWS
and
L. T. FAN

College of Engineering,
Kansas State University
Manhattan, Kansas 66506

A semifluidized bed combines the advantages of both packed and fluidized beds. This paper compares the performance of semifluidized beds and packed beds in the adsorption of organic compounds from water, for various flow rates and fixed-bed to fluidized-bed height ratios. It was found that the same amount of adsorption occurs in semifluidized beds at a lower pressure drop than in fixed beds.

INTRODUCTION

Adsorption is an important unit in process in the purification of gases and liquids and in bulk separations in the chemical process industries. It is also widely used in water purification, wastewater renovation, and the control of toxic chemicals in the environment. The adsorptive transfer of materials from the fluid phase to the solid phase is effected in batch slurry adsorbers, continuous flow stirred tank reactors (CSTR) or in columnar operations. Batch adsorbers are principally used in kinetic and equilibrium studies and in small volume processing applications. Examples of CSTR type operation are in water treatment applications for the removal of taste and odor compounds and in wastewater treatment using the PACT[R] process (Hutton and Temple, 1979). By far the most common form of contacting, in practice, is in columns of fixed, fluidized or moving sorbent beds.

Fixed bed adsorbers are used in a large number of industrial applications due to the ease and reliability of operation of these process units. Mass transfer rates are relatively high in fixed-beds, but pressure drop considerations limit the use of small sized particles in these beds. Countercurrent operation, with the solids moving countercurrent to the fluid is one of the most efficient methods of contacting. However, movement of large amounts of solids can add to the operating problems and also may result in the loss of friable adsorbents due to attrition.

Countercurrent operation is approximated in practice by having several fixed-beds in series, pulsed-bed operation, or by the Simulated Moving Bed Process. Discussions on these and other types of process arrangements for adsorption are available in the literature (Broughton, 1977; Hutchins, 1980; Vermeulen, 1977).

Fluidized bed operation allows the use of relatively small particles, assures a uniform operating temperature and also provides a lower column pressure drop compared to packed beds. This type of operation is also advantageous when the processing fluids contain large amounts of suspended solids that may easily clog a packed bed. The use of single and multi-stage fluidized beds for adsorption applications has been reported by Parmele et al. (1979) and Kunii and Levenspiel (1969).

The effects of combining two types of reactor systems, such as a CSTR followed by a tubular reactor (MT) or a tubular reactor followed by a CSTR (TM), have been studied theoretically for isothermal reactors by Cholette et al. (1960). The degree of conversion is a maximum for the tubular reactor alone, and decreases as the fraction of fluid processed in the CSTR is increased, for first and higher order reactions. Cholette and Blanchet (1961) and Cholette and Coutier (1959) have shown from theoretical analysis that for endothermal reactions, the tubular reactor is always superior to a CSTR, while for exothermal reactions the CSTR is superior to the tubular reactor up to a certain conversion after which the tubular reactor is more efficient. For an exothermal adiabatic reaction, the advantages of MT type of operation have been reported by Aris (1962) and Douglas (1964). For reactions approaching zero order, the degree of conversion is independent of the reactor geometry.

Semifluidized beds combine the features of both fluidized beds and packed beds in a single reactor. Application of semifluidized bed reactors for chemical processing, biochemical processing and solid-liquid separations have been studied extensively in the past. Mass transfer and pressure drop data for solid-liquid systems and gas-solid systems were reported by Fan et al. (1960), and Wen et al. (1963). The performances of tubular, mixed and semifluidized bed reactors in the oxidation of benzene were compared by Rao and Doraiswamy (1970). The semifluidized bed reactor was reported to have a definite advantage over the other reactors. Fan and Hsu (1980) have evaluated the usefulness of semifluidized beds in biochemical processing, particularly in the waste treatment area. Hsu and Fan (1982) have noted substantial improvements in the filter run time when semifluidized beds were used in the filtration of coal slurries.

The objectives of this study were to evaluate experimentally the performance of semifluidized-bed adsorbers, and to compare it with packed-bed and fluidized-bed adsorbers. Results reported here are limited to the removal of trace organics from aqueous solutions by adsorption on activated carbon. Desorption and pressure drop data are also reported for the different modes of adsorber operation.

EXPERIMENTAL METHODS

Reagent grade phenol was used as the solute for all the experiments. Two types of bituminous coal-based activated carbon, Filtrasorb-400 (F-400) and APC, supplied by the Calgon Corporation were used as adsorbents. Both contain particles in the 12 to 40 sieve size range. The F-400 carbon has surface area in the range of 1000-1200 m^2/gm, a pore volume of 0.94 ml/gm; the particle density ranges between 1.3 and 1.4 gm/ml. The APC carbon has a surface area of 1525 m^2/gm, a pore volume of 1.2 ml/gm and a particle density of 2.25 gm/ml. The carbons were sieved, washed with deionized distilled water and dried to constant weight at 110°C. The 18/20 size fraction (passing No. 18 sieve and retained on No. 20 sieve) was used in all studies.

A schematic of the experimental system used is shown in Figure 1. All of the experiments were conducted in a plexiglass column 5.08 cm in diameter. The column had an adjustable sieve plate that could be used to operate the bed as a packed, semifluidized or completely fluidized bed. Glass beads 2mm in diameter were used at the bottom of the column for flow distribution. The water was supplied to the column by a positive displacement pump fed from a constant head tank filled with tap water filtered through a preadsorption column. A metering pump discharged a concentrated solution of phenol into the water line. Influent flow rates and concentrations were monitored to assure constant input conditions. Effluent concentrations were measured to obtain the breakthrough profiles.

The adsorption column was operated as a packed, semifluidized or fluidized bed by positioning the sieve plate in the column. For fluidized bed operation the sieve plate was moved to the top allowing free movement of solids in the bed. Incipient fluidization occurred at a flow rate of 0.18 m^3/m^2min (i.e. m/min), and the bed was vigorously fluidized at a flow rate of 0.32 m/min. For semifluidized bed operation, the fluid bed was compressed by moving the screen down to form two distinct sections of packed and fluidized beds. By positioning the sieve plate at various heights, different ratios of packed and fluid bed heights were obtained. The bed was completely compressed allowing no bed expansion, when the bed was operated as a packed bed.

RESULTS AND DISCUSSION

The experimental conditions for adsorption using both the F-400 and APC carbons are shown in Table 1. All tests were run with 648 grams of 18/20 mesh size carbon having a geometric mean diameter of 917 μm. The packed bed heights for all the runs using the same type of carbon were approximately the same, assuring uniform packing in the columns.

Experimental breakthrough profiles for packed, semifluidized and fluidized beds are shown in Figure 2 for a flow rate of 0.408 m/min. The breakthrough concentration at which the adsorption operation is to be terminated was arbitrarily selected to be 5% of the influent concentration, and comparisons were made between the time of operation before regeneration for the three different modes of operation. The breakthrough time was 5.5 hours for the fluidized bed, 20 hours for the semifluidized bed with 60% fluidization, and 22 hours for the packed bed. The effect of semifluidization is evident in the dramatic increase in the breakthrough time, from 5.5 hours for the fluidized bed to 20 hours for the semifluidized bed with 60% fluidization.

The breakthrough curves at the hydraulic loading rate of 0.815 m/min for the packed and semifluidized beds are shown in Figures 3 and 4. The breakthrough times for the effluent to reach 5% of the influent concentration level were 7.2 hours for the packed bed, 6.9 hours for the semifluidized bed with 23% fluidization, 6.6 hours for that with 50% fluidization and 6.0 hours for that with 75% fluidization. The breakthrough profiles are close to each other, indicating that approximately the same amount of adsorption can be achieved in a semifluidized bed as in a packed bed under identical loading conditions.

The breakthrough curves for APC carbon at a flow rate of 0.571 m/min shown in Figure 5 indicates results similar to those noted above. The time for the effluent concentration to reach 5% of the influent level was approximately the same for the packed bed and for the semifluidized bed with 25% or 50% fluidization. The breakthrough time is 9.5 hours for the packed bed, 9.1 hours for the semifluidized bed with 25% fluidization bed and 9.1 hours for the semifluidized bed with 50% fluidization.

Pressure Drop and Economic Comparisons

Pressure drop data for F-400 and APC carbons are shown for a range of flow rates and various fractions of bed fluidization in Figures 6 and 7, respectively. Column pressure drop decreased as the percentage of bed that is fluidized was increased. A comparison is shown in Table 2 for the performance of packed beds and that of semifluidized beds based on the volume of fluid processed per unit pressure drop. The volume of fluid processed ΔF, was obtained in each case by multiplying the breakthrough time at $C/C_o = 0.05$ with the volumetric flow rate. As evident from Table 2, the volume of fluid processed per unit pressure drop ($\Delta F/(\Delta P/L)$) increased as the fraction of the bed that is fluidized was increased. The value of $\Delta F/(\Delta P/L)$ was maximum for the semifluidized bed and fell to lower values for either packed or fluidized bed operation. The bed fraction fluidized in the semifluidized bed adsorber to produce maximum economic advantage will vary as a function of the operating variables and the adsorbate and adsorbent properties. These conditions need to be determined by experimental and mathematical analyses. Also, in addition to $\Delta F/(\Delta P/L)$, the adsorbent regeneration frequency must be considered in the economic analysis. Thus, even though the fluidized bed has a higher $\Delta F/(\Delta P/L)$ compared to the packed bed at the hydraulic loading rate of 0.408 m/min, the fluidized bed does not appear to be economically advantageous due to the more rapid breakthrough, and hence higher regeneration frequency. This is not the case with the semifluidized bed adsorber, since the time for breakthrough for the semifluidized bed is approximately the same as that for the packed bed.

The economic advantage of utilizing semifluidized beds for practical adsorption applications will depend on the volume of fluid being processed. The larger the volume of fluid processed and the deeper the bed, the greater will be the savings in power cost. For example, a 380,000 m^3/d (100 mgd) plant processing water can reduce annual pumping cost by approximately $11,500/year per 0.3m (1 ft.) of head saved, assuming 7¢/KWH for power and 70% efficiency for the pumping system.

The total dynamic head required for the three different modes of operation may not be substantially different from each other. This is due to the fact there is an increase in static head requirement as the fraction of bed that is fluidized is increased. Although this head must be available at the upstream end of the adsorber, this head is not lost and is available for downstreams use.

Experimental Results from Desorption Studies

Desorption studies were conducted at a flow rate of 0.4 m/min with the F-400 carbon and at a flow rate of 0.571 m/min with the APC carbon. Desorption was carried out with tap water without any solute, and the effluent concentrations were monitored. The initial concentration at which the columns were switched from adsorption to desorption was 9.16×10^{-4} M for the fluidized bed, 9.04×10^{-4} M for the semifluidized bed and

8.85 x 10^{-4} M for the completely packed bed. For the APC carbon, the initial concentration was 9.38 x 10^{-4} M for the semifluidized bed and 9.08 x 10^{-4} M for the packed bed.

Desorption data for the F-400 carbon are shown in Figure 8. The fluidized bed had the slowest rate of desorption, while the rates for the semifluidized and packed beds were comparable. The effluent concentration was 5% of the initial concentration at 37.5 hours for the fluidized bed, 30.5 hours for the semifluidized bed, and 27.5 hours for the packed bed.

The desorption data for the APC carbon are shown in Figure 9. They indicate a more rapid desorption rate with the semifluidized bed. The semifluidized bed effluent concentration reached 5% of the initial concentration at 26 hours compared to 35 hours for the packed bed. These results indicate that cyclic adsorption-desorption operations may be feasible at a lower overall pressure drop using semifluidized beds in the cases where the sorbent beds are regenerated using solvents.

CONCLUSIONS

Three types of adsorber operation, namely fluidized bed, semifluidized bed and packed bed have been compared for the removal of trace concentrations of phenol from aqueous solutions. Breakthrough occurs fairly rapidly with the fluidized bed, and at comparable times for the semifluidized and packed beds. The volume of fluid that can be processed for the same amount of pressure drop is much higher for the semifluidized bed than the packed bed. Desorption data indicate essentially similar behavior for the F-400 carbon. For the APC carbon desorption rate appears to be faster in the semifluidized bed than in the packed bed.

Semifluidized beds can be easily adapted for use in adsorption and ion-exchange applications. Water and wastewater processing operations potentially can realize significant savings by using this mode of operation because of the large volumes of fluid processed. However, additional experimental research and modeling need to be performed to optimize the system with respect to a variety of single solutes and multi-solute systems. Moreover, the effect of variabilities in flow and concentration on the performance need to be evaluated. Preliminary results from additional studies being conducted by us indicate that in addition to pressure drop savings, we can prolong appreciably the breakthrough time of the semifluidized bed over that of the packed bed.

LITERATURE CITED

1. Aris, R. (1962). Can. J. Chem. Engg., $\underline{40}$, 39.

2. Broughton, D. B. (1977). Chem. Engg. Progr., $\underline{73}$ (11), 49.

3. Cholette, A., and J. Blanchet (1961). Can. J. Chem. Engg., $\underline{39}$, 192.

4. Cholette, A., J. Blanchet and L. Cloutier (1960). Can. J. Chem. Engg., $\underline{38}$, 1.

5. Cholette, A., and L. Coutier (1959). Can. J. Chem. Engg., $\underline{37}$, 105.

6. Douglas, J. M. (1964). Chem. Engg. Progr. Symp. Ser., $\underline{61}$ (48), 1.

7. Fan, L. T., Y. C. Yang and C. Y. Wen (1960). AIChE J., $\underline{6}$, 482.

8. Fan, L. T. and E. H. Hsu (1980). "Semifluidized Bed Bioreactor". Paper presented at the VIth International Fermentation Symposium, London, Ontario, Canada, July 20-25.

9. Hsu, E. H. and L. T. Fan (1982). "A Novel Porous Medium Filter", Proceedings of the World Filtration Contress III, pp. 420-425, Philadelphia, PA, September. 13-17.

10. Hutchins, R. A. (1980). Chem. Engg., $\underline{87}$ (4), 101.

11. Hutton, D. G. and S. Temple (1979). "Priority Pollutant Removal: Comparison of Du Pont PACT Process and Activated Sludge", Paper presented at the 52nd Annual WPCF Convention, Houston, Texas.

12. Kunii, D. and O. Levenspiel (1969). "Fluidization Engineering", John Wiley and Sons, Inc., New York.

13. Parmele, C. S., W. L. O'Connel and H. S. Basdekis (1979). Chem. Engg., $\underline{86}$ (28), 59.

14. Rao, B. K. and L. K. Doraiswamy (1970). AIChE J., 16, 273.

15. Vermeulen, T. (1977). Chem. Engg. Progr., 73 (11), 57.

16. Wen, C. Y., S. C. Wang and L. T. Fan (1963). AIChE J., 9, 316.

Table 1. Experimental Conditions for Column Studies*

Type of Carbon	Hydraulic Loading Rate (m/min)	Height of Fixed Bed (cm)	Height of Fluidized Bed (cm)	% of Bed Fluidized
Calgon F-400	0.408	78.6	0.0	0.0
		38.1	55.8	59.4
		0.0	114.3	100.0
Calgon F-400	0.815	66.0	0.0	0.0
		58.8	18.0	23.4
		44.2	43.8	49.8
		29.0	74.4	75.5
Calgon APC	0.571	81.4	0.0	0.0
		69.8	24.0	25.6
		55.8	55.8	50.0

* Weight of carbon is 647.8 gms for all runs.

Table 2. Comparison of Performance of Packed and Semifluidized Bed Adsorbers

Hydraulic Loading Rate (m/min)	Fraction of Bed Fluidized	Pressure Drop Per Unit Depth $\Delta P/L$	Volume Processed at $C/C_0 = 0.05$ $\Delta F (m^3)$	$\Delta F/(\Delta P/L)$ (m^3)	$\frac{(\Delta F/(\Delta P/L))_{SBA}}{(\Delta F/(\Delta P/L))_{PBA}}$
0.408*	0.000	0.650	1.09	1.68	1.00
	0.594	0.350	0.99	2.83	1.69
	1.000	0.122	0.27	2.23	1.33
0.815*	0.000	1.330	0.50	0.38	1.00
	0.234	1.120	0.48	0.43	1.14
	0.498	0.825	0.46	0.56	1.48
	0.755	0.567	0.42	0.74	1.96
0.571**	0.000	1.400	0.94	0.67	1.00
	0.256	0.960	0.90	0.94	1.40
	0.500	0.700	0.90	1.29	1.92

* F-400 Carbon
** APC Carbon

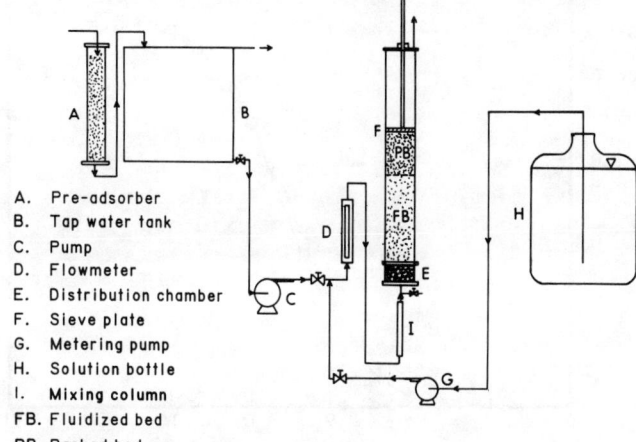

A. Pre-adsorber
B. Tap water tank
C. Pump
D. Flowmeter
E. Distribution chamber
F. Sieve plate
G. Metering pump
H. Solution bottle
I. Mixing column
FB. Fluidized bed
PB. Packed bed

Figure 1. Schematic of adsorption system.

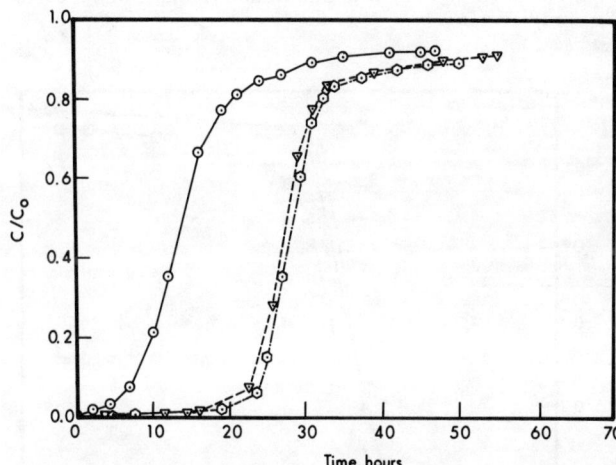

Figure 2. Breakthrough profiles for (○) fluidized-bed (▽) semifluidized bed with 59% fluidization and (⊙) packed bed adsorbed (F-400 carbon, 0.408 m³/m²min flow rate).

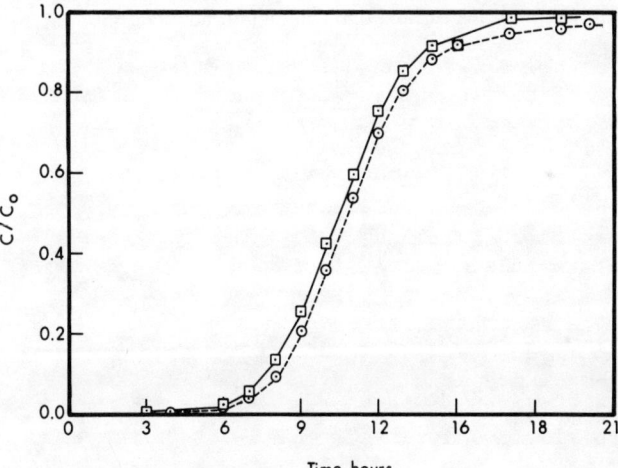

Figure 3. Breakthrough profiles for (○) packed bed and (□) semifluidized bed with 23% fluidization (F-400 carbon, 0.815 m³/m²min flow rate).

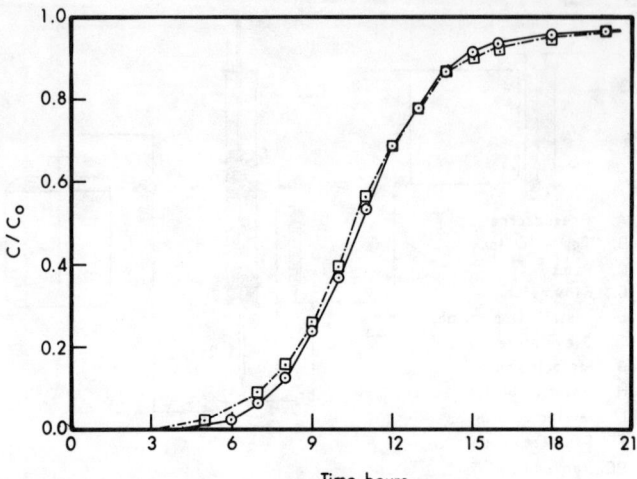

Figure 4. Breakthrough profiles for semifluidized beds with (○) 50% fluidization and (□) 75% fluidization (F-400 carbon, 0.815 m³/m²min flow rate).

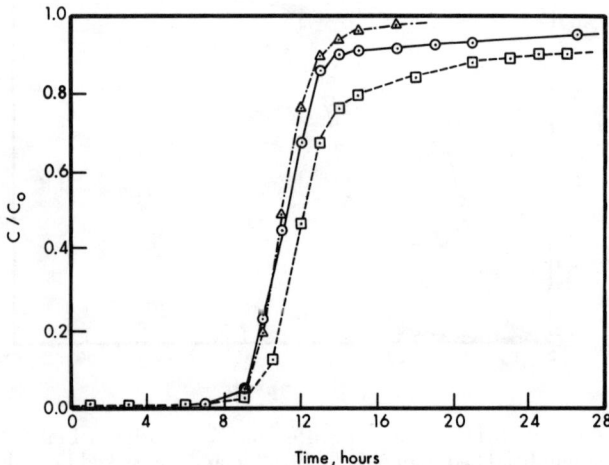

Figure 5. Breakthrough profiles for (□) packed bed and semifluidized beds with (△) 25% fluidization and (○) 50% fluidization (APC carbon, 0.571 m³/m²min flow rate).

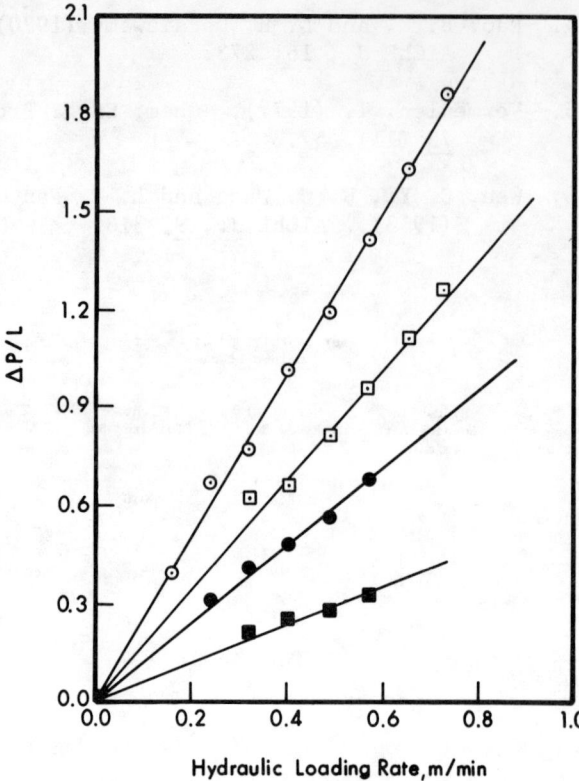

Figure 6. Pressure drop data for F-400 carbon: (○) packed bed and semifluidized beds with (□) 25% fluidization, (●) 50% fluidization and (■) 75% fluidization.

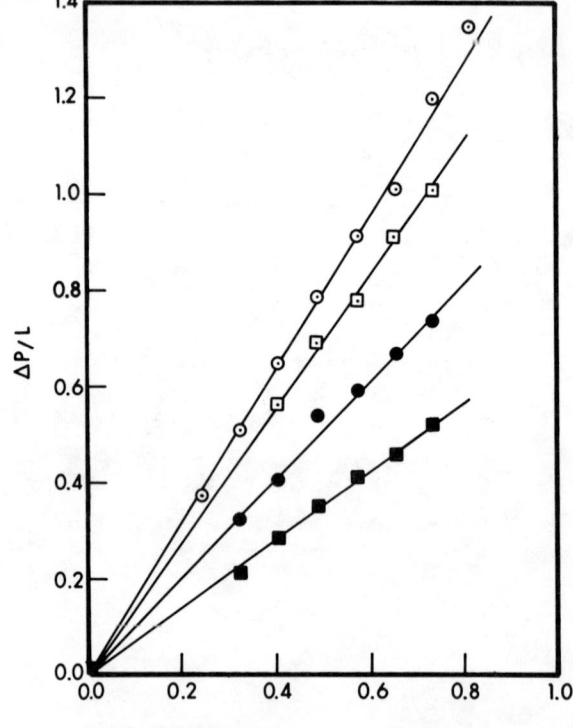

Figure 7. Pressure drop data for APC carbon: (○) packed bed and semifluidized beds with (□) 25% fluidization, (●) 50% fluidization and (■) 75% fluidization.

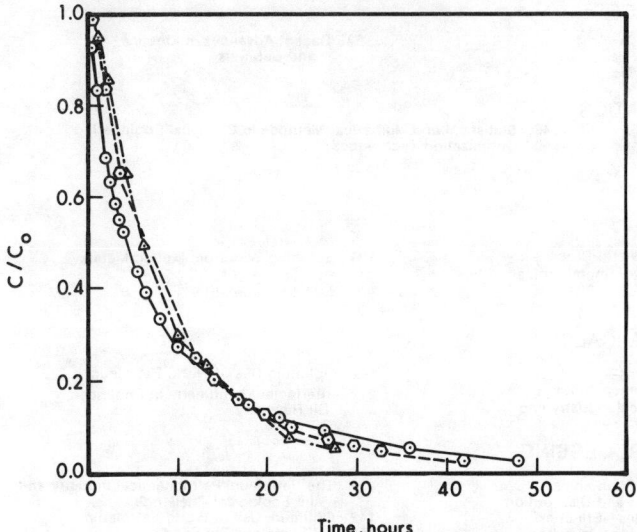

Figure 8. Desorption curves for (○) fluidized bed, (◯) semifluidized bed with 59% fluidization and (△) packed bed adsorbers (F-400 carbon, 0.408 m³/m²min flow rate)

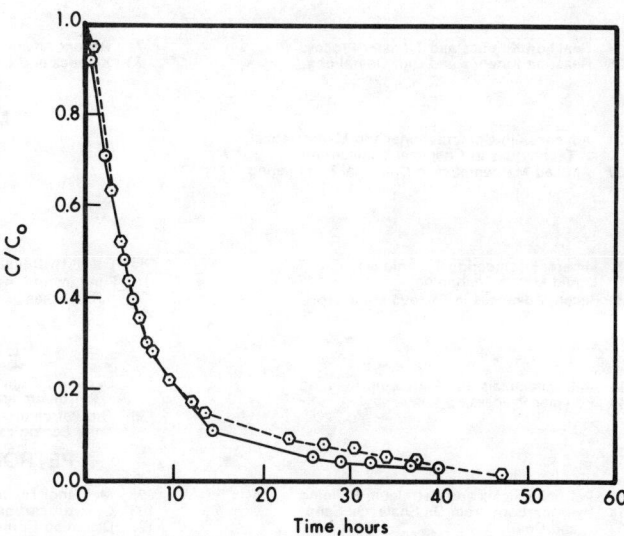

Figure 9. Desorption curves for (◯) packed bed (○) semifluidized bed with 50% fluidization (APC carbon, 0.571 m³/m²min flow rate).

KINETICS

- 4 Reaction Kinetics and Transfer Process
- 25 Reaction Kinetics and Unit Operations
- 72 Recent Advances in Kinetics
- 73 Kinetics and Catalysis
- 83 Recent Advances in Kinetics and Catalysis

MATHEMATICS

- 31 Advances in Computational and Mathematical Techniques in Chemical Engineering
- 37 Applied Mathematics in Chemical Engineering
- 42 Statistical and Numerical Methods in Chemical Engineering
- 50 Optimization Techniques

MINERALS

- 15 Mineral Engineering Techniques
- 20 Liquid Metals Technology—Part I
- 43 Recent Advances in Ferrous Metallurgy
- 85 Fossil Hydrocarbon and Mineral Processing
- 173 Fundamental Aspects of Hydrometallurgical Processes
- 180 Spinning Wire from Molten Metals
- 216 Processing of Energy and Metallic Minerals

PETROCHEMICALS

- 34 Petrochemicals and Petroleum Refining
- 49 Polymer Processing
- 127 Declining Domestic Reserves—Effect on Petroleum and Petrochemical Industry
- 135 The Petroleum/Petrochemical Industry and the Ecological Challenge
- 142 Optimum Use of World Petroleum
- 212 Interfacial Phenomena In Enhanced Oil Recovery

PETROLEUM PROCESSING

- 34 Petrochemicals and Petroleum Refining
- 54 Hydrocarbons from Oil Shale, Oil Sands, and Coal
- 98 Methanol Technology and Economics
- 103 C_4 Hydrocarbon Production and Distribution
- 127 Declining Domestic Reserves—Effect on Petroleum and Petrochemical Industry
- 135 The Petroleum/Petrochemical Industry and the Ecological Challenge
- 142 Optimum Use of World Petroleum
- 155 Oil Shale and Tar Sands

PHASE EQUILIBRIA

- 2 Pittsburgh and Houston
- 3 Minneapolis and Columbus
- 6 Collected Research Papers
- 81 Phase Equilibria and Related Properties
- 88 Phase Equilibria and Gas Mixtures Properties

PROCESS DYNAMICS

- 36 Process Dynamics and Control
- 46 Process Systems Engineering
- 55 Process Control and Applied Mathematics
- 159 Chemical Process Control
- 214 Selected Topics on Computer-Aided Process Design and Analysis

SEPARATION

- 91 Unusual Methods of Separation
- 120 Recent Advances in Separation Techniques
- 192 Recent Advances in Separation Techniques—II

SONICS

- 1 Ultrasonics—Two Symposia
- 109 Sonochemical Engineering

THERMODYNAMICS

- 7 Applied Thermodynamics
- 44 Thermodynamics
- 140 Thermodynamics—Data and Correlations

TRANSPORT PROPERTIES

- 16 Mass Transfer
- 56 Selected Topics in Transport Phenomena
- 77 Fundamental Research on Heat and Mass Transfer

MISCELLANEOUS

- 26 Chemical Engineering Education—Academic and Industrial
- 48 Chemical Engineering Reviews
- 70 Small-Scale Equipment for Chemical Engineering Laboratories
- 76 High Pressure Technology
- 112 Engineering, Chemistry, and Use of Plasma Reactors
- 125 Vacuum Technology at Low Temperatures
- 143 Standardization of Catalyst Test Methods
- 160 Continuous Polymerization Reactors
- 182 Biorheology
- 183 The Modern Undergraduate Laboratory Innovative Techniques
- 185 Electro organic Synthesis Technology
- 186 Plasma Chemical Processing
- 187 Chronic Replacement of Kidney Function
- 194 Hazardous Chemical—Spills and Waterborne Transportation
- 203 A Review of AIChE's Design Institute for Physical Property Data (DIPPR) and Worldwide Affiliated Activities
- 204 Tutorial Lectures in Electrochemical Engineering and Technology
- 206 Controlled Release Systems
- 224 Cryogenic Processes and Equipment 1982

MONOGRAPH SERIES

- 1 Reaction Kinetics by Olaf Hougen
- 2 Atomization and Spray Drying by W. R. Marshall, Jr.
- 3 The Manufacture of Nitric Acid by the Oxidation of Ammonia—The DuPont Pressure Process by Thomas H. Chilton
- 4 Experiences and Experiments with Process Dynamics by Joel O. Hougen
- 5 Present, Past, and Future Property Estimation Techniques by Robert C. Reid
- 6 Catalysts and Reactors by James Wei
- 7 The 'Calculated' Loss-of-Coolant Accident by L. J. Ybarrondo, C. W. Solbrig, H. S. Isbin
- 8 Understanding and Conceiving Chemical Process by C. Judson King
- 9 Ecosystem Technology: Theory and Practice by Aaron J. Teller
- 10 Fundamentals of Fire and Explosion by Daniel R. Stull
- 11 Lumps, Models and Kinetics in Practice by Vern W. Weekman, Jr.
- 12 Lectures in Atmospheric Chemistry by John H. Seinfeld
- 13 Advanced Process Engineering by James R. Fair
- 14 Synfuels from Coal by Bernard S. Lee